W9-BBP-045

Almost Everywhere Convergence II

Proceedings of the International Conference on
Almost Everywhere Convergence in
Probability and Ergodic Theory
Evanston, Illinois
October 16–20, 1989

Almost Everywhere Convergence II

Proceedings of the International Conference on Almost Everywhere Convergence in Probability and Ergodic Theory
Evanston, Illinois
October 16–20, 1989

Edited by

Alexandra Bellow
Department of Mathematics
Northwestern University
Evanston, Illinois

Roger L. Jones
Department of Mathematics
DePaul University
Chicago, Illinois

ACADEMIC PRESS, INC.
Harcourt Brace Jovanovich, Publishers
Boston San Diego New York
London Sydney Tokyo Toronto

This book is printed on acid-free paper. ∞

ACADEMIC PRESS, INC.
1250 Sixth Avenue, San Diego, CA 92101

United Kingdom Edition published by
ACADEMIC PRESS LIMITED
24–28 Oval Road, London NW1 7DX

Library of Congress Cataloging-in-Publication Data

International Conference on Almost Everywhere Convergence in Probability and Ergodic Theory (2nd : 1989 : Evanston, Ill.)
Almost everywhere convergence II : proceedings of the International Conference on Almost Everywhere Convergence in Probability and Ergodic Theory, Evanston, Illinois. October 16–20, 1989 / edited by Alexandra Bellow, Roger Jones.
p. cm.
Includes bibliographical references.
ISBN 0-12-085520-8 (alk. paper)
1. Convergence — Congresses. 2. Inequalities (Mathematics) — — Congresses. 3. Probabilities — Congresses. 4. Ergodic theory — — Congresses. I. Bellow, A. (Alexandra), 1935 — . II. Jones, Roger (Roger L.) III. Title.
QA295.I56 1989
519.2—dc20 91-19343
CIP

Printed in the United States of America
91 92 93 94 9 8 7 6 5 4 3 2 1

CONTENTS

CONTRIBUTORS

M. A. Akcoglu, *Department of Mathematics, University of Toronto, Toronto, Ontario M5S 1A1, Canada*

Idris Assani, *Department of Mathematics, CB#3250, Phillips Hall, University of North Carolina, Chapel Hill, North Carolina, 27599*

Alexandra Bellow, *Department of Mathematics, Northwestern University, Evanston, Illinois, 60601*

Erich Berger, *Institut für Mathematische Stochastik, Universität Göttingen, Lotzestr. 13, D-3400 Göttingen, Germany*

N. H. Bingham, *Department of Mathematics, Royal Holloway & Bedford New College, Egham Hill, Egham, Surrey TW20 OEX, England*

Robert Bradley, *Department of Mathematics, Northwestern University, Evanston, Illinois, 60601*

M. Broise, *Département de Mathématiques et Informatique, Université de Bretagne Occidentale, 6, Avenue Victor Le Gorgeu, 29287 Brest, France*

Alberto P. Calderón, *Department of Mathematics, University of Chicago, Chicago, Illinois, 60637*

Doğan Çömez, *Department of Mathematics, North Dakota State University, Fargo, North Dakota, 58105*

A. del Junco, *Department of Mathematics, University of Toronto, Toronto, Ontario M5S 1A1, Canada*

Y. Déniel, *Département de Mathématiques et Informatique, Université de Bretagne Occidentale, 6, Avenue Victor Le Gorgeu, 29287 Brest, France*

Yves Derriennic, *Département de Mathématiques et Informatique, Université de Bretagne Occidentale, 6, Avenue Victor Le Gorgeu, 29287 Brest, France*

Roger L. Jones, *Department of Mathematics, DePaul University, 2219 N. Kenmore, Chicago, Illinois, 60614*

John C. Kieffer, *Electrical Engineering Department, University of Minnesota, Twin Cities, 200 Union Street SE, Minneapolis, Minnesota, 55455*

I. Kornfeld, *Department of Mathematics, North Dakota State University, Fargo, North Dakota, 58105*

Ulrich Krengel, *Department of Mathematical Stochastics, University of Göttingen, Lotzestrasse 13, 3400 Göttingen, Germany*

Tze Leung Lai, *Department of Statistics, Stanford University, Stanford, California, 94305*

W. M. F. Lee, *Department of Mathematics, University of Toronto, Toronto, Ontario M5S 1A1, Canada*

Michael Lin, *Department of Mathematics, Ben–Gurion University of the Negev, Bèer Sheva, Israel*

James Olsen, *Department of Mathematics, North Dakota State University, Fargo, North Dakota, 58105*

Magda Peligrad, *Department of Mathematical Sciences, University of Cincinnati, Cincinnati, Ohio 45221*

Karl Petersen, *Department of Mathematics, CBℓ3250, Phillips Hall, University of North Carolina, Chapel Hill, North Carolina, 27599*

Pál Révész, *Institute of Statistics, Technical University of Vienna, Wiedner Hauptstrasse 8–10/107, A–1040 Vienna, Austria*

L. C. G. Rogers, *School of Mathematical Sciences, Queen Mary & Westfield College, Mile End Road, London, E1 4NS, England*

Joseph Rosenblatt, *Department of Mathematics, The Ohio State University, Columbus, Ohio, 43210*

Yoram Sagher, *Department of Mathematics, University of Illinois – Chicago, Chicago, Illinois 60680*

Louis Sucheston, *Department of Mathematics, The Ohio State University, Columbus, Ohio, 43210*

László I. Szabó, *Department of Mathematics, József Attila University, Bolyai Institute, Po. box:656, Aradi vértanúk tere 1, H–6720 Szeged, Hungary*

Homer White, *Department of Mathematics, CBℓ3250, Phillips Hall, University of North Carolina, Chapel Hill, North Carolina, 27599*

Rainer Wittmann, *Department of Mathematical Stochastics, University of Göttingen, Lotzestrasse 13, 3400 Göttingen, Germany*

Kecheng Zhou, *Department of Mathematics, California State University – Sacramento, Sacramento, California 95819*

CONFERENCE PARTICIPANTS

Idris Assani
Rodrigo Banuelos
Alexandra Bellow
Vitaly Bergelson
Erich Berger
N. H. Bingham
Daniel Boivin
Michael Boshernitzan
Jean Bourgain
Robert Bradley
Donald Burkholder
Alberto Calderón
Robert Cogburn
Doḡan Çömez
Burgess Davis
Victor de la Peña
Alberto de la Torre
Yves Derriennic
Aryeh Dvoretzky
G. A. Edgar
Robert Fefferman
Albert Fisher
Shaul Foguel
H. Furstenberg
Vladimir Gaposhkin
Constantine Georgakis
Pawel Hitzcenko
William B. Johnson
Roger L. Jones
Shizuo Kakutani
Y. Katznelson
John C. Kieffer
I. Kornfeld
Ulrich Krengel
James Kuelbs
Michael Lacey
Tze Leung Lai
Ehud Lehrer
Michael Lin
Michael Marcus
F. J. Martin–Reyes
Terry McConnell
James Olsen
Steven Orey
Donald Ornstein
Magda Peligrad
Karl Petersen
Mark Pinsky
M. B. Rao
Karin Reinhold–Larsson
Pál Révész
Joseph Rosenblatt
Yoram Sagher
Elias M. Stein
Louis Sucheston
Arkady A. Tempelman
Mate Wierdl
Rainer Wittmann
Wojbor A. Woyczynski
Radu Zaharopol

This volume is dedicated to the memory of Steven Orey

As these proceedings were going to press we received the news of the tragic death of Steven Orey, one of the organizers of the Conference.

A creative and broad mathematician he made important contributions to Logic and Probability. He was also a dedicated and inspiring teacher, who contributed to the growth of mathematics by training many gifted PhD students. His kind manner and passion for mathematics, his intellectual sophistication and refinement, coupled with his modesty and moral integrity, attracted not only students but other mathematicians and colleagues as well.

Steven Orey was the main founder of the important Probability center at the University of Minnesota.

He was also our friend and with many others, we feel his loss deeply.

Preface

The Second International Conference on Almost Everywhere Convergence in Probability and Ergodic Theory took place at Northwestern University on October 16–20, 1989. Financial Support for the conference was provided by the National Science Foundation, the National Security Agency, and the Institute for Mathematics and Applications, as well as by Northwestern University. The conference was in some ways a continuation of the very successful conference which Louis Sucheston organized at Ohio State University in June, 1988.

There were approximately 60 conference participants. In addition to mathematicians from all over the United States, conference participants included mathematicians from Austria, Germany, France, Great Britain, Israel, the Soviet Union, and Spain.

The "local organizers" of the conference were Alexandra Bellow and Mark Pinsky of Northwestern University and Roger Jones of DePaul University. The organizing committee also included Donald Burkholder, Gerald Edgar, Donald Ornstein, Steven Orey and Louis Sucheston.

In recent years there have been many remarkable developments in almost everywhere convergence. This work was reflected in the number and variety of talks given by outstanding mathematicians. The papers in these proceedings are an expansion of many of the important talks at the conference. For example, Karl Petersen discussed the helical transform defined by

$$H_\theta f(x) = \lim_{n\to\infty} \sum_{\substack{k=-n \\ k\neq 0}}^{n} \frac{e^{ik\theta} f(T^k x)}{k}$$

and the relationship it has with important operators in harmonic analysis, such as the partial sum operator, which plays a crucial role in proving almost everywhere convergence of Fourier series. In 1971 Krengel showed that there is a sequence $\{n_k\}$ such that for any ergodic measure preserving point transformation T on a non–atomic probability space, there is a function $f \in L^1$ for which

$$\lim_{N\to\infty} \frac{1}{N} \sum_{k=1}^{N} f(T^{n_k} x)$$

fails to exist a.e. Such sequences are called "bad universal". Subsequently it was shown by A. Bellow that general lacunary sequences are "bad universal" in L^1, and later (with R. Jones and V. Losert) that the ergodic averages along lacunary sequences of the form a^k, $a\geq 2$, have the "strong sweeping out property". In his talk Joseph Rosenblatt gave a very different and clever argument, using Jean Bourgain's powerful entropy method, to show that general lacunary sequences and sequences such as $\{k\,[\log k]\}$ are "bad universal", not only for L^1, but for L^∞. Ulrich Krengel discussed Hopf's ergodic theorem for particles with different velocities, and described the solution to a problem that had been posed by Hopf in 1936. In his talk Louis Sucheston extended some of his very nice earlier work on almost everywhere convergence of processes indexed by directed sets. John Kieffer defined the notion of a log–convex set of random variables, and proved a general almost sure convergence theorem for sequences of log–convex sets. This theorem contains the subadditive ergodic theorem of Kingman and the Shannon–McMillan– Breiman Theorem. Yves Derriennic discussed maximal inequalities and rearrangements, again showing the connections between harmonic analysis and ergodic theory. In his talk Pál Révész discussed a generalization of the almost sure central limit theorem as it relates to logarithmic density. Many other interesting and stimulating talks were also given.

In addition to talks related to papers in these proceedings, there were other important talks at the conference. Jean Bourgain discussed the Riesz Raikov Theorem, both the progress that has been made, and the open problems that remain. Vladimir Gaposhkin of the Soviet Union talked about his work on the use of spectral methods to obtain results on a.e. convergence. Robert Fefferman and Elias Stein discussed results in harmonic analysis that are related to a.e. convergence. There were also important talks by Vitaly Bergelson, Burgess Davis, Hillel Furstenberg, Shizuo Kakutani, Donald Ornstein, Arkadi Tempelman, and many others.

A Solution to a Problem of A. Bellow*

M.A. Akcoglu A. del Junco W.M.F. Lee

The following problem was posed by A. Bellow in [1]. Let $\{t_n\}$ be a sequence of real numbers converging to 0. Let f be an integrable function of a real variable. Is it then true that the sequence

$$f_n(x) = \frac{1}{n}\sum_{k=1}^{n} f(x + t_k)$$

converges a.e. to f ? In [2] J. Bourgain gave a negative answer to this question, even when f is also a bounded function, as an application of a deep general theorem proved in the same article. Our purpose in this note is to obtain an elementary and self contained solution to this problem. Our method is an extension of a method of W. Rudin [3], incorporating an idea from [2]. We are grateful to Karl Petersen for very helpful discussions on this subject.

It will be more convenient to work with the interval $X = [0, 1)$, rather than the real line R. We consider X with its Lebesgue measure λ and with the metric $d(x, y) = |e^{2\pi ix} - e^{2\pi iy}|$, which makes X a compact metric space. In what follows $\{t_k\}$ denotes a sequence of non-zero numbers in X, converging to 0. For any $t \in X$ we let δ_t denote the unit point mass at t, and define, for $n \geq 1$,

$$\mu_n = \frac{1}{n}\sum_{k=1}^{n} \delta_{t_k}.$$

Our purpose is to prove the following result.

Theorem *For any $\varepsilon > 0$ there is a set $C \subset X$ such that $\lambda(C) < \varepsilon$ and such that*

$$\limsup_n (\chi_C * \mu_n) > 1/8$$

*Research supported in part by NSERC Grants

ISBN 0-12-085520-8

λ-a.e. on X.

Here χ_C denotes the characteristic function of C and $f * \mu$ is the convolution of a function $f : X \to \mathrm{R}$ with a measure μ on X. The proof of this theorem will follow from a series of lemmas.

For any $\alpha \in \mathrm{R}$ we define $R_\alpha : X \to X$ and $P_\alpha : X \to X$ by $R_\alpha(x) = (\alpha + x)\mathrm{mod}1$ and $P_\alpha(x) = (\alpha x)\mathrm{mod}1$, $x \in X$. Note that R_α is a measure-preserving isometry for all $\alpha \in \mathrm{R}$. If m is a positive integer then P_m is a measure-preserving mapping. We will consider P_m only for positive integer values of m. We observe that $\lambda(A \cap P_m^{-1}B)$ converges to $\lambda(A)\lambda(B)$ as $m \to \infty$, for any (measurable) $A, B \subset X$. If $0 < \varepsilon < 1/2$ then we let $G_\varepsilon = [\varepsilon, 1-\varepsilon] \subset X$ and $H_\varepsilon = G_\varepsilon^c = X - G_\varepsilon$. Note that H_ε is an open set containing 0.

Lemma 1. *For every integer $n \geq 1$ there are infinitely many integers m such that $\mu_n(P_m^{-1}G_{1/6}) > 3/8$.*

Proof If χ is the characteristic function of $G_{1/6}$ and if $t \in X$, $t \neq 0$, then one can easily verify that

$$\lim_M \frac{1}{M} \sum_{m=1}^{M} \chi(P_m(t)) \geq 1/2.$$

This means that

$$\lim_M \frac{1}{M} \sum_{m=1}^{M} \delta_t(P_m^{-1}G_{1/6}) \geq 1/2.$$

Since each μ_n is a convex combination of finitely many unit point masses, the same result is also true if δ_t is replaced by μ_n. Then the lemma follows.

Lemma 2. *Let Q be an isometry of a compact metric space Y, H an open subset of Y, and $y_0 \in H$. Then there is a number K and an infinite sequence of integers $\{k_i\}$ such that, for all j, $0 < k_{j+1} - k_j < K$, and $Q^{k_j}y_0 \in H$.*

We omit the easy and well-known proof.

Lemma 3. *Let F be a finite subset of X, $N_0 \geq 1$, and let $0 < \varepsilon < 1/2$. Then there is an integer $n_0 \geq N_0$ and infinitely many integers m such that*

$$P_m F \subset H_\varepsilon,$$

$$\mu_{n_0}(P_m^{-1} G_{1/8}) > 1/4.$$

Proof Let $F = \{s_1, \ldots, s_a\}$ and let $Y = X^a$ be the Cartesian product of a copies of X. Let $Q : Y \to Y$ be defined by

$$Q(x_1, \ldots, x_a) = (R_{s_1}(x_1), \ldots, R_{s_a}(x_a)).$$

Then Q is an isometry of a compact metric space. Since H_ε^a is an open set in Y containing the point $O = (0, \ldots, 0)$, Lemma 2 shows the existence of a K and a sequence $\{k_j\}$ such that $0 < k_{j+1} - k_j < K$ and $Q^{k_j} O \in H_\varepsilon^a$. Since $Q^k O = (P_k s_1, \ldots, P_k s_a)$, we see that this last condition is equivalent to $P_{k_j} F \subset H_\varepsilon$ for all j. Now find an N_1 such that $K d(t_n, 0) < 1/100$ for all $n \geq N_1$. Then let $n_0 = \max\{N_0,\ 100 N_1\}$. By Lemma 1 there are infinitely many integers m such that $\mu_{n_0}(P_m^{-1} G_{1/6}) > 3/8$. For each one of these integers m let I_m be the set of integers n such that $N_1 \leq n \leq n_0$ and $P_m(t_n) \in G_{1/6}$. Then we see that there are more than $n_0/4$ integers in I_m. If k is another integer such that $|m - k| < K$, then we also see that $P_k(t_n) \in G_{1/8}$ for all $n \in I_m$. In particular we can choose k as one of the terms in the sequence $\{k_j\}$ obtained above. Hence the conditions of the Lemma are satisfied for an infinite subsequence of $\{k_j\}$.

The first two conclusions of the following lemma correspond to a simpler version of Lemma 11 of [2].

Lemma 4. *Let $0 < \xi$, $0 < \eta$, $0 < \varepsilon < 1/2$. Let $N \geq 1$ and $K \geq 1$ be two integers and let E_n^k, $1 \leq n \leq N$, $1 \leq k \leq K$ be KN sets in X. Then there are $2N$ integers $1 \leq n_1 < \ldots < n_N$ and $1 \leq m_1 < \ldots < m_N$, such that*

$$\mu_{n_i}(P_{m_j}^{-1} H_\varepsilon) > 1 - \eta, \tag{1}$$

if $i \neq j$,

$$\mu_{n_i}(P_{m_i}^{-1} G_{1/8}) > 1/4, \tag{2}$$

and

$$| \lambda(\bigcap_{i=1}^{N}(P_{m_i}^{-1}E_i^k)) - \prod_{i=1}^{N}\lambda(E_i^k) |< \xi \tag{3}$$

for $1 \leq k \leq K$.

Proof We apply an induction on N. If $N = 1$ then we obtain n_1 and m_1 from Lemma 3, by letting the finite set F in that lemma be the empty set. In fact in this simple case one can choose $n_1 \geq 1$ arbitrarily. Now assume that $N > 1$ and that $n_1 < \ldots < n_{N-1}$ and $m_1 < \ldots < m_{N-1}$ are already chosen to satisfy the conditions (1), (2), and (3) with $N-1$ instead of N. We find an N_1 such that $m_{N-1}d(t_n, 0) < (1/10)\varepsilon$ for all $n \geq N_1$. Then we choose N_0 such that $\eta N_0 > N_1$. This guarantees that if $n \geq N_0$ then $\mu_n(P_{m_i}^{-1}H_\varepsilon) > 1 - \eta$ for all i, $1 \leq i \leq N-1$. We then apply Lemma 3, with this choice of N_0, with the given ε, and with the finite set $F = \{t_1, \ldots, t_{n_{N-1}}\}$, to obtain n_N and infinitely many possible choices for m_N that would satisfy the conditions (1) and (2). It is clear that if m_N is chosen sufficiently large then (3) will also be satisfied.

Lemma 5. *Let* $0 < \xi$, $0 < \eta$, $0 < \zeta$, $0 < \varepsilon$, $\zeta + \varepsilon < 1/2$, *and let* $N \geq 1$ *be an integer. Then there are* $2N$ *integers* $1 \leq n_1 < \ldots < n_N$ *and* $1 \leq m_1 < \ldots < m_N$, *such that, if*

$$C = \bigcap\{P_{m_i}^{-1}G_\zeta \mid 1 \leq i \leq N\ \} \tag{4}$$

and

$$A_j = \bigcap\{P_{m_i}^{-1}G_{\zeta+\varepsilon} \mid 1 \leq i \leq N, i \neq j\ \}\bigcap P_{m_j}^{-1}H_\zeta, \tag{5}$$

$1 \leq j \leq N$, *then*

$$\mu_{n_i}(P_{m_j}^{-1}H_\varepsilon) > 1 - \eta, \tag{6}$$

if $i \neq j$,

$$\mu_{n_i}(P_{m_i}^{-1}G_{1/8}) > 1/4, \tag{7}$$

and

$$| \lambda(C) - (1 - 2\zeta)^N |< \xi, \tag{8}$$

$$| \lambda(A_j) - 2\zeta(1 - 2\zeta - 2\varepsilon)^{N-1} |< \xi \tag{9}$$

for $1 \leq j \leq N$.

Proof Apply Lemma 4 with $K = N + 1$ and with an obvious choice for the sets E_n^k.

Lemma 6. *With the notations of Lemma 5, if* $0 < \zeta + \varepsilon < 1/100$ *and if the conditions (6) and (7) are satisfied then*

$$(\chi_C * \mu_{n_j})(x) > 1/4 - (N-1)\eta$$

for all $x \in A_j$, $1 \le j \le N$.

Proof Let $1 \le j \le N$ and $x \in A_j$. We will estimate the number of points $x_k = (x - t_k)\text{mod}1$, $1 \le k \le n_j$, that are contained in each one of the sets $P_{m_i}^{-1} G_\zeta$, $1 \le i \le N$. This is the same as estimating the number of points

$$P_{m_i}(x_k) = (P_{m_i}(x) - P_{m_i}(t_k))\text{mod}1$$

that are contained in G_ζ. If $i = j$ then $P_{m_j}(x) \in H_\zeta$ and more than 1/4 of the points $P_{m_j}(t_k)$, $1 \le k \le n_j$, are in $G_{1/8}$. Hence more that 1/4 of the points $P_{m_j}(x_k)$, $1 \le k \le n_j$, are in $G_{1/8-\zeta} \subset G_\zeta$. If $i \ne j$ then $P_{m_i}(x) \in G_{\zeta+\varepsilon}$ and more than $(1-\eta)$-portion of the points $P_{m_i}(t_k)$, $1 \le k \le n_j$, are in H_ε. Hence more than $(1-\eta)$-portion of the points $P_{m_i}(x_k)$, $1 \le k \le n_j$, are in G_ζ. Hence we see that more than $(1/4 - (N-1)\eta)$-portion of the points x_k, $1 \le k \le n_j$, are in C, which is the conclusion of the lemma.

Lemma 7. *Given positive integers* R *and* s *there are sets* B *and* C *in* X *such that*

$$0 < R\lambda(C) < \lambda(B) \tag{10}$$

and such that

$$\sup_{s<n}(\chi_C * \mu_n)(x) > 1/8 \tag{11}$$

for all $x \in B$.

Proof Choose a small ζ, for example $\zeta = 1/1000$. Then choose an integer N such that

$$\frac{(N-s)2\zeta}{1-2\zeta} > 2R,$$

and take $\eta = 1/(10N)$. Then find $\varepsilon > 0$ such that

$$(1 - \frac{2\varepsilon}{1-2\zeta})^{N-1} > \frac{3}{4}.$$

Hence, with these choices,

$$\frac{(N-s)2\zeta(1-2\zeta-2\varepsilon)^{N-1}}{(1-2\zeta)^N} > \frac{3}{2}R.$$

Finally take a $\xi > 0$ such that

$$\frac{(N-s)2\zeta(1-2\zeta-2\varepsilon)^{N-1}-(N-s)\xi}{(1-2\zeta)^N+\xi} > R.$$

We then apply Lemma 5 to obtain C, A_j, and then let $B = \cup_{s<j\leq N} A_j$. Since the sets A_j are pairwise disjoint,

$$\lambda(B) > (N-s)2\zeta(1-2\zeta-2\varepsilon)^{N-1}-(N-s)\xi,$$

and, of course,

$$\lambda(C) < (1-2\zeta)^N+\xi.$$

We then see that (10) is satisfied. Also, because of Lemma 6,

$$1/8 < 1/4-(N-1)\eta < \sup_{s<j\leq N}(\chi_C * \mu_{n_j})(x) \leq \sup_{s<n}(\chi_C * \mu_n)(x),$$

for all $x \in B$, since $j \leq n_j$.

Proof of the main result It is essentially known that Lemma 7 implies the Theorem stated at the beginning. We will briefly sketch the argument for completeness. First note that if the sets B and C satisfy the conclusions of Lemma 7, then the sets $R_\alpha B$ and $R_\alpha C$ will also satisfy these conditions. Now it is clear that Lemma 7 implies the existence of two sequences of sets $\{C_j\}$ and $\{B_j\}$, and a squence of integers $\{s_j\}$, converging to infinity, such that $\sum_j \lambda(C_j) < \infty$, $\sum_j \lambda(B_j) = \infty$, and such that

$$\sup_{s_j<n}(\chi_{C_j} * \mu_n)(x) > 1/8$$

for all $x \in B_j$. Note that repetitions are allowed in all these sequences. For a given $\varepsilon > 0$ we may also assume that $\sum_j \lambda(C_j) < \varepsilon$. Using a standard argument, stated as Lemma 8 below, we find a sequence of real numbers $\{\alpha_j\}$ such that λ-almost all points in X belong to infinitely many of the sets $R_{\alpha_j}B_j$. We then see that $C = \cup_j R_{\alpha_j} C_j$ satisfies the requirements of the Theorem.

Lemma 8 *Let $\{B_j\}$ be a sequence of sets in X such that $\sum_j \lambda(B_j) = \infty$. Then there is a sequence of real numbers $\{\alpha_j\}$ such that*

$$\lambda(\cup_{n \le j} R_{\alpha_j} B_j) = 1$$

for all n.

We will omit the standard proof, which uses the fact that given any $M > 1$ and two sets E and F with positive measure there is an α such that

$$\lambda(E \cap R_\alpha F) < M\lambda(E)\lambda(F).$$

It is clear that a similar result is true for any ergodic flow.

References

1. A. Bellow, *Two problems*, Proc. Oberwolfach Conference on Measure Theory (June 1981), Springer Lecture Notes in Math. **945**, (1982).

2. J. Bourgain, *Almost sure convergence and bounded entropy*, Israel J. Math. **63**, (1988), 79-97.

3. W. Rudin, *An arithmetic property of Riemann sums*, Proc. AMS **15**, (1964), 321-324.

Department of Mathematics
University of Toronto
Toronto, Ontario M5S 1A1, Canada

Universal Weights from Dynamical Systems To Mean–Bounded Positive Operators on L^p

Idris Assani

Abstract:

Let (a_n), $a_n \in \ell_+^\infty$ be a universal weight for all dynamical systems (i.e., $\frac{1}{n}\sum_{k=0}^{n-1} a_k \cdot f(\varphi^k(\omega))$ converges a.e. for all $f \in L^\infty(\mu) \cap L^1(\mu)$ and all dynamical systems $(\Omega, a, \mu, \varphi)$ where φ is an invertible measure preserving transformation on (Ω, a, μ)). We show that (a_n) is also a universal weight for all mean bounded positive operators on L^p i.e., $\frac{1}{n}\sum_{k=0}^{n-1} a_k T^k f(\omega)$ converges a.e. for all positive mean bounded operators T on L^p, $1<p<\infty$ (p is fixed).

[1]Department of Mathematics, Phillips Hall
University of North Carolina at Chapel Hill
Chapel Hill, NC 27599

Research supported in part by University Research Council Grant 5–4427

ISBN 0-12-085520-8

Introduction

Classical ergodic theorems deal mainly with Cesaro averages. Recently averages taken along subsequences and weighted averages have received considerable attention. The present article deals with this type of averages in the setting of mean bounded positive operators.

Definition.

Let (a_n) be a sequence of complex numbers. We say that (a_n) is a universal weight (for the pointwise ergodic theorem, P.E.T.) if for all dynamical systems (Ω,a,μ,S) and for $f \in L^\infty(\mu)\cap L^1(\mu)$ the sequence $\frac{1}{n}\sum_{k=0}^{n-1} a_k f(S^k\omega)$ converges a.e.

J. Bourgain, H. Furstenberg, Y. Katznelson and D. Ornstein [2], proved that universal weight could be found if neglecting single null set we consider the sequence $a_k = g(\varphi^k x)$ where $g \in L^\infty$ and $(X,\mathcal{F},\nu,\varphi)$ is a dynamical system.

In [1] we proved that their result could be extended to the setting of mean bounded positive (m.b.p.) operators on L^p, $1<p<\infty$ (p is fixed). (T is mean bounded if $\sup_n \| M_n(T) \|_p < \infty$; $M_n(T) = \frac{I+T+\cdots+T^{n-1}}{n}$). Namely the sequence $\frac{1}{n}\sum_{k=0}^{n-1} a_k \cdot T^k f(w)$ converges a.e. for all $f \in L^p$, p fixed, $1<p<\infty$ and all m.b.p. operators on L^p, where $a_k = g(\varphi^k x)$. We also proved that m.b.p. operators could create some new weights for by the following result.

Theorem A [1]:

Let T be a m.b.p. operator on $L^p(\mu)$ and $g \in L^p(\mu)$, p fixed, $1<p<\infty$. Then the sequence $((T^k g)(x))_k$ is a.e. a universal weight.

An interesting consequence of this result is that m.b.p. operators satisfy the Wiener–Wintner property: we can find a single null set off which $\frac{1}{n}\sum_{k=0}^{n-1} (T^k g)(x)\, e^{2\pi i k\epsilon}$

converges for all $\epsilon \in \mathbb{R}$. (The null set of x is the same for all $\epsilon \in \mathbb{R}$.)

We want in this note to show that weighted ergodic theorems for dynamical systems can be lifted to the setting of m.b.p. operators. This is a partial answer to a more general problem; having a pointwise ergodic result for dynamical systems (subsequence of integers) do we have the corresponding result for m.b.p. operators on L^p?

Our main result is the following

Theorem 1.

Let (a_n) be a universal weight, $(a_n) \in \ell^\infty_+$. Then for all m.b.p. operators T on $L^p(\Omega,a,\mu)$, p fixed, $1<p<\infty$ ((Ω,a,μ) can be assumed to be a Lebesgue space hence isomorphic to the unit interval with Lebesgue measure) and all $f \in L^p$ the sequence $\frac{1}{n}\sum_{k=0}^{n-1} a_k \cdot T^k f(\omega)$ converges a.e.

The proof follows the same ideas as in [1]. By taking $a_k = g(\varphi^k x)$, $g \in L^\infty_+$ we can easily get the first extension obtained in [1] of the result of J. Bourgain, H. Furstenberg, Y. Katznelson and D. Ornstein. We will need for the proof of Theorem 1 the following two propositions. (In Proposition 2, supp Inv T is the maximal support of the invariant functions for T.) For more details on previous results on universal weights for dynamical systems and operators we refer to [1a] and [1b].

Proposition 2. [1]

Let T be a m.b.p. operator on $L^p(Y,G,\nu)$, $1<p<\infty$, p fixed. Then there exists a decomposition of the space in three disjoint sets C_1, D_1, Z and a set $C_2 \subset Z$ with the following properties

1) $\text{supp Inv } T = C_1 \cup C_2$

2) $1_{D_1} \cdot \sum_{n=0}^{\infty} T^n g < \infty$ a.e. for all $g \in L^p(\nu)$

3) $M_n(T)(1_Z \cdot g) \to 0$ a.e. for all $g \in L^p(\nu)$

4) $\operatorname{supp} \operatorname{Inv} T^* = C_1 \cup D_1$

5) $1_{C_2} \sum_{n=0}^{\infty} T^{*n} h < \infty$ a.e. for all $h \in L^q(\nu)$, $(\frac{1}{p} + \frac{1}{q} = 1)$

6) $M_n(T^*)\left[1_{D_1 \cup Z} \cdot h\right] \xrightarrow[n]{} 0$ a.e.

Proposition 3. [1]

Let (a_n) be a universal weight, $(a_n) \in \ell^\infty$ and T a m.b.p. operator on $L^p(Y,G,\nu)$, $1<p<\infty$, p fixed. Then for all $g \in L^p(\nu)$ the sequences

i) $$1_{C_1} \frac{1}{n} \cdot \sum_{k=0}^{n-1} a_k T^k(1_{C_1} \cdot g)$$

ii) $$1_{D_1 \cup Z} \frac{1}{n} \cdot \sum_{k=0}^{n-1} a_k T^k g$$

converge a.e.

Proof of Theorem 1. In view of these two propositions it remains only to show that the sequence $1_{C_1} \frac{1}{n} \cdot \sum_{k=0}^{n-1} a_k T^k(1_{D_1} g)(y)$ converges a.e. (ν). We proceed in several steps.

Step 1: We can reduce the question to the study of the convergence of $1_{C_1} \frac{1}{n} \cdot \sum_{k=0}^{n-1} a_k U^k(1_{D_1} g)$ where U is a positive contraction on $L^1(\bar{C})$ ($\bar{C}$ =supp Inv T^*). To justify this let us take $\nu_0^* \in L^q(\nu)$ such that $T^*(\nu_0^*) = \nu_0^*$, and $\operatorname{supp} \nu_0^* = \bar{C}$. Then define U on $L^1\left[\bar{C}, \bar{C} \cap G, \nu_{|\bar{C}}\right]$ by $U(\nu_0^* \cdot g) = \nu_0^* \cdot Tg$. The operator U extends to a positive linear contraction on $L^1(\bar{C})$ as

$$\int U(\nu_0^* \cdot g)\, d\nu = \int \nu_0^* \cdot Tg \cdot d\nu = \int \nu_0^* \cdot g \cdot d\nu.$$

As noted in [1], U satisfies also the mean ergodic theorem in $L^1(\bar{C})$ and hence U satisfies also the P.E.T. on $L^1(\bar{C})$. Thus by E. M. Nikishin's theorem [5] we have the following maximal inequality.

$$m\left\{y \in C_1 : \sup_n \left|\frac{1}{n}\sum_{k=0}^{n-1} U^k(1_{C_1}\cdot g)(y)\right| > \lambda\right\} \le \frac{K}{\lambda}\|g\|_1$$

where m is a measure equivalent to $\nu|_{C_1 \cup D_1}$. In particular we have

$$m\left\{y \in C_1 : \sup_n \left|\frac{1}{n}\sum_{k=0}^{n-1} a_k U^k(1_{C_1}g)(y)\right| > \lambda\right\} \le \frac{K}{\lambda}\|a\|_\infty \cdot \|g\|_1 \, .$$

Step 2. In this step and the next ones we assume that $g \geq 0$, $g \in L^1$.

We can easily check that

$$1_{C_1}\cdot \frac{1}{n}\sum_{k=0}^{n-1} a_k \cdot U^k(1_{D_1}g)(y) \le 1_{C_1}\frac{1}{n}\sum_{k=0}^{n-1} U^k\left[\sum_{j=0}^{\infty} 1_{C_1}\cdot a_{k+j}\left[(U\,1_{D_1})^j g\right]\right](y)\, .$$

Step 3. We want to prove that the sequence

$$1_{C_1}\cdot \frac{1}{n}\sum_{k=0}^{n-1} U^k\left[\sum_{j=0}^{\infty} 1_{C_1}a_{k+j}\left[(U\,1_{D_1})^j g\right]\right](y) \text{ converges a.e.}$$

To show this we can apply Proposition 3 (i) for each j to $\left[1_{C_1}a_{k+j}(U\,1_{D_1})^j g\right]$. (Remark that if $(a_k)$is a universal weight then (a_{k+j}) is also a universal weight). Having the result for each integer M for the sequence

$$\frac{1}{n}\sum_{k=0}^{n-1} 1_{C_1}\left[U^k\left[\sum_{j=0}^{M} 1_{C_1} a_{k+j}\left[(U\,1_{D_1})^j g\right]\right](y)\right]$$

the conclusion follows by using the weak type inequality and the facts (see [4]) that

$$H_{C_1}(g)(y) = \sum_{j=0}^{\infty} 1_{C_1}\left[U(1_{D_1})^j(g)\right](y) < \infty .$$

and

$$\int H_{C_1}(g)\, d\nu \leq \int g\, d\nu .$$

Step 4. We want to prove that

$$\underline{\lim}\, 1_{C_1} \cdot \frac{1}{n}\sum_{k=0}^{n-1} a_K\left[U^k(1_{D_1} g)\right](y)$$

$$\geq \lim_n 1_{C_1} \frac{1}{n}\sum_{k=0}^{n-1} U^k\left[\sum_{j=0}^{\infty} 1_{C_1} a_{k+j}\left[(U\,1_{D_1})^j(g)\right]\right](y) .$$

To show this we are going to use the filling scheme for the positive contraction $\bar{U} = \phi \times U$ on $(\mathbb{Z} \times (C_1 U D_1))$ where ϕ is the shift on $\mathbb{Z}$ with the counting measure. So if $\bar{g}(k,y) = b(k)\cdot(g(y)$ then $(\bar{U}\bar{g})(k,y) = b(k+1)\cdot(Ug)(y)$. We notice also that if

$\bar{g}(k,y) = b_1(k)\cdot g_2(y) + a(k)\cdot h_2(y)$ and
$\bar{g}^1(k,y) = b_1(k)\cdot g_2(y) + a(k+1)(Uh_2)(y)$ then

$$\frac{1}{n}\sum_{k=0}^{n-1} (\bar{U}^{k'}\bar{g})(k,y) = \frac{1}{n}\left[\sum_{k'=0}^{n-1} (\bar{U}^{k'}\bar{g}^1)(k,y)\right] - \frac{a(k+n)\cdot(U^n h_2)(y)}{n} + \frac{a(k)\; h_2(y)}{n}$$

And so

$$\frac{1}{n}\sum_{k'=0}^{n-1}(\bar{U}^{k'}\tilde{g})(k,y) - \frac{1}{n}\sum_{k'=0}^{n-1}(\bar{U}^{k'}\tilde{g}^{1})(k,y) \longrightarrow 0 \text{ a.e. } (\nu).$$

as $\left[\frac{U^n(h_2))(y)}{n}\right] \to 0$ a.e. (ν).

If we write $\tilde{g} \xrightarrow{1} \tilde{g}^1$ as in [4, p. 120] and denote by $\bar{U}_{\mathbb{Z}\times C_1}$ the operator $1_{\mathbb{Z}\times C_1} + \bar{U}\left[1_{\mathbb{Z}\times D_1}\right]$ we have (see [4, p. 127])

$$\bar{U}^m_{\mathbb{Z}\times C_1}(1_{\mathbb{Z}\times D_1}\, a(k)\, g(y)) \xrightarrow[m]{\nearrow} 1_{C_1}\sum_{j=0}^{\infty} a(j)\left[(U\, 1_{D_1})^j(g)\right](y)$$

and

$$1_{\mathbb{Z}\times D_1} a(k) g(y) \xrightarrow{m} \bar{U}^m_{\mathbb{Z}\times C_1}(1_{\mathbb{Z}\times D_1}\, a(k)\, g(y))\,.$$

(The $\xrightarrow{m}$ refers to the filling scheme.) By now a simple induction on m (see [4] for instance for more details on this argument) and the property shown for $m=1$ we can deduce that

$$\underline{\lim}\, M_n(\bar{U})\left[1_{\mathbb{Z}\times D_1}(a(k))\cdot(g)\right] \geq \underline{\lim}_n\, M_n(\bar{U})\left[1_{C_1}\sum_{j=0}^{\infty} a(j)(U\, 1_{D_1})^j g\right](y)$$

$$= \underline{\lim}_n \frac{1}{n}\sum_{k=0}^{n-1} U^k \left[\sum_{j=0}^{\infty} a(k+j)\left[(U\, 1_{D_1})^j(g)\right](y)\right]$$

$$= \lim_n \text{———————————————}$$

(by Step 3)

It is easy now to obtain the desired conclusion by comparing step 2 and step 4.

Remarks and Questions.

1) It is easy to see that if R is an operator such that $R^n f(\omega)$ converges a.e. then $(R^n f(\omega))$ is a universal weight.

An operator with such a property is the operator A used in Brunel's proof; starting with an m.b.p. operator T on $L^p(\mu)$ we can get an operator $A(T) =$ (formally) $\frac{1}{T}(I - \sqrt{I-T})$ such that

$$M_n(T)f \leq \gamma M_{[\sqrt{n}]}(A)f \text{ for all } f \geq 0 \quad (\text{see [2]})$$

for this operator we have $\underline{\underline{A}}^n f$ converges a.e. for $f \in L^p(\mu)$. It seems interesting to know if by just using the connections between A and T a proof of Theorem 1 could be obtained without using the result in [2]. This would simplify at least the content of these arguments.

2) It is possible to combine Theorems A and 1 to obtain universality from m.b.p. operators in $L^p(\mu)$ to m.b.p. operators in $L^q(\mu)$, for instance.

References

1. I. Assani, "The return times and the Wiener–Wintner property for mean bounded positive operators in Lp," submitted.

1a. J. R. Baxter and J. H. Olsen, "Weighted and subsequential ergodic theorems," Canad. J. Math. 35 (1983), 145–168.

1b. A. Bellow and V. Losert, "The weighted pointwise ergodic theorem and the individual ergodic theorem along subsequences," Trans. A. M. S. 288, no. 1, (1985), 307–346.

2. J. Bourgain, H. Furstenberg, Y. Katznelson, and D. Ornstein, "Return time sequences," Appendix to J. Bourgain: "Pointwise ergodic theorems on arithmetic sets," I.H.E.S. preprint (1989).

3. A. Brunel, "Le theoreme ergodique ponctuelle pour les operateurs positifs à moyennes bornees dans Lp," preprint.

4. U. Krengel, Ergodic Theorems, de Gruyter, 1985.

5. E. M. Nikishin, "Resonance theorems and superlinear operators," translation of Uspekhi, Mat. Nauk., vol. XXV, no. 6, Nov.–Dec. 1970.

SOME CONNECTIONS BETWEEN ERGODIC THEORY AND HARMONIC ANALYSIS

Idris Assani[1], Karl Petersen[2], and Homer White

1. Introduction.

Parallels, connections, and cross-applications among ergodic theory, harmonic analysis, and spectral theory have long been observed and exploited. We focus on a family of operators that exists in the boundary zone of these three areas, the study of which clarifies some of these interelationships: the "rotated ergodic Hilbert transform", or, more briefly, the *helical transform* (of an integrable function f on a measure space X with respect to a measure-preserving transformation T):

$$H_\theta f(x) = \lim_{n \to \infty} \sum_{k=-n}^{n}{}' \frac{e^{ik\theta} f(T^k x)}{k} .$$

For a fixed θ the existence of this limit for almost every x is a direct

[1]Research supported in part by Univ. Res. Council Grant 5-4427.

[2]Research supported in part by NSF Grant DMS-8900136.

ISBN 0-12-085520-8

consequence of the existence of the ordinary ergodic Hilbert transform [8, 6, 18]. There is also an analogous class of operators for a measure-preserving flow $\{T_t\colon -\infty<t<\infty\}$. Consideration of the special cases of translation on **Z** or **R** leads directly into harmonic analysis; indeed, a relative of the rotated Hilbert transform ((8) below) played a key role in Carleson's proof [7] of the a.e. convergence of the Fourier series of L^2 functions. The interest in the helical transform in ergodic theory rose originally from its connection with spectral theory [5]; but to prove for measure-preserving transformations a strengthening of Gaposhkin's necessary and sufficient spectral condition for a unitary operator to satisfy the pointwise ergodic theorem [11] it became necessary to apply the Carleson-Hunt Theorem to establish a double maximal inequality (taking suprema over both n and θ) for the helical transform. This implication can also be reversed, so it appears that almost everywhere convergence results in harmonic analysis and ergodic theory are perhaps even more closely linked than has been realized beforehand. It is possible that further study of these relationships may lead to improvements in some of these results such as simpler proofs or maybe even slightly stronger statements.

We discuss the relationships among strong (p, p) inequalities for the maximal helical transform and its variants and the suprema of the partial sums of Fourier series. For the case $p=2$, the Carleson-Hunt estimate is equivalent to the L^2 boundedness of the maximal helical transform. For $p\neq 2$, many implications among these maximal inequalities still hold, but certain ones fail. Weak (1, 1) also fails for the double maximal helical transform, as does the Wiener-Wintner property. The L^2 boundedness of the double maximal helical transform extends to measure-preserving flows as well as to higher-dimensional actions. This opens the way for applications such as a strengthening, for measure-preserving flows, of Gaposhkin's necessary and sufficient spectral condition for a flow of unitary operators to satisfy the Local Ergodic Theorem [12]. These results are proved in [3, 1, 2], so we omit most of the arguments, except that the (p, p) case we treat in more detail. The equivalence between the lemma on partial Fourier coefficients [5] and the Carleson-Hunt Theorem when $p=2$ is mirrored in that between the weakening of the partial Fourier coefficients lemma that arises by restricting the stopping points to 2^{-r}, $r\geq 0$, and Kolmogorov's theorem on the a.e. convergence of the 2^r'th partial sums of the Fourier series of an L^2 function. We give a proof of the partial Fourier coefficients lemma in this restricted form by means of kernel estimates and Fourier transforms in the style of Gaposhkin [11] and Bourgain [4]. Then Kolmogorov's theorem follows quickly, by the same path that leads from the full partial Fourier coefficients lemma to the Carleson-Hunt Theorem; we give this proof in detail. Finally, we show how these same easy estimates allow one to prove a quadratic-variation version of the Hardy-Littlewood maximal lemma on l^2 similar to Bourgain's estimate along the sequence of squares, and consequently a quadratic-variation strengthening of Birkhoff's Pointwise Ergodic Theorem. This latter point is implicit in [4]; many of these expository matters are treated in [19].

2. Equivalences and implications among several maximal estimates.

We will say that an operator $M{:}\,L^p(X)\to L^p(Y)$, where X and Y are measure spaces, is *strong type* (p, p) in case there is a constant C such that $\|Mf\|_p \leq C\|f\|_p$ for all $f\in L^p(X)$. Usually **R** is taken together with Lebesgue measure, [0, 1] with normalized Lebesgue measure, and **Z** or any of its subsets with counting measure, in which case we denote L^p by l^p. We follow the ancient convention, going back to freshman calculus, of denoting by C various constants, not necessarily all the same. The $'$ on summations means that terms with zero denominator are to be omitted.

We are interested in relations among the strong (p, p) property for the following operators, among others.

(1) Maximal operator for partial sums of Fourier series (Carleson-Hunt [7, 14]). For a function $\hat{h}$ on $\mathbf{Z}$ and $x \in [0, 1)$, let

$$S^{*}\hat{h}(x) = \sup_{n>0} \left| \sum_{j=-n}^{n} \hat{h}(j)\, e^{2\pi i j x} \right| .$$

(2) Maximal range for a helical walk. For $N=1, 2, \ldots$ and $r=1, \ldots, N$ let $\omega_r = e^{2\pi i r/N}$. For a function v on $\{1, \ldots, N\}$ define

$$W^{*}v(r) = \sup_{1 \leq M \leq N} \left| \frac{1}{\sqrt{N}} \sum_{m=1}^{M} v(m)\omega_r{}^{m} \right| .$$

(3) Maximal partial Fourier coefficients [5]. For $h \in L^{p}[0, 1)$ and $j \in \mathbf{Z}$, let

$$\Gamma^{*}h(j) = \sup_{t>0} \left| \int_0^t h(x)\, e^{-2\pi i j x}\, dx \right| .$$

(4) Maximal Helical Transform on l^p. For $a \in l^p(\mathbf{Z})$ define

$$H^{*}a(j) = \sup_{\theta \in \mathbf{R}} \left| \sum_{k=-\infty}^{\infty}{}' \frac{e^{2\pi i (j-k)\theta}\, a_{j-k}}{k} \right| .$$

THEOREM 1 [3]: The strong type (2, 2) property for each of the operators in (1)-(4) is formally equivalent to strong type (2, 2) for each of the others.

Sketch of the proof. The equivalence between strong (p, p) for (3) and (4) is clear upon computing the partial Fourier coefficients of a function $h(x)=\sum_{k=-\infty}^{\infty} a_k e^{2\pi ikx}$ in $L^2[0, 1)$:

$$\int_0^{\epsilon} h(x)\, e^{-2\pi ijx}\, dx = a_j\epsilon + \sum_{k=-\infty}^{\infty}{}' \frac{e^{2\pi i(k-j)\epsilon}\, a_k}{k-j} - \sum_{k=-\infty}^{\infty}{}' \frac{a_k}{k-j} .$$

The third term on the right is the discrete Hilbert transform, which is well known (and easily shown by taking Fourier transforms) to be strong type (2, 2). Moreover, since the first and third terms on the right are actually strong type (p, p) for all $p>1$, it follows that H^* on l^p is strong type (p, p) if and only if the same is true of

$$\widetilde{T}^* a(j) = \sup_{\epsilon>0} \left| \int_0^{\epsilon} \Big(\sum_{k=-\infty}^{\infty} a_k e^{2\pi ijx}\Big)\, e^{-2\pi ijx}\, dx \right| .$$

The equivalence between strong (2, 2) for (2) and (3) is demonstrated by applying (3) to step functions. That strong (2, 2) for (2) implies strong (2, 2) for (1) follows by considering Riemann sum approximations to the integrals involved. The details are given in [3] and also below, for the special case of lacunary stopping times, so we turn now to the interesting path from strong (2, 2) for (1) to strong (2, 2) for (2).

Fix $N=1, 2, \ldots$ and let $\alpha=1/(2N)$. We use an argument of Davenport and Halberstam [9] involving "well-spaced points" $x_r=r/N$ for $r=1, \ldots, N$, and a "well-chosen function"

$$\psi(x) = \sum_{n=-\infty}^{\infty} \left[\frac{\sin(n\pi\alpha)}{n\pi\alpha}\right]^2 e^{2\pi inx} .$$

The key elements are the product formula for the Fourier transform of a convolution, Hölder's inequality, and the fact that intervals of radius α

about points 2α apart are disjoint.

If $\gamma_\alpha = (\chi_{[1-\alpha/2,\ 1]} + \chi_{[0,\ \alpha/2]})/\alpha$, then $\psi = \gamma_\alpha * \gamma_\alpha$, so that

$$\psi(x) = \begin{cases} \frac{1}{\alpha}\left(1 - \frac{|x|}{\alpha}\right) & \text{if } |x| \le \alpha \\ 0 & \text{if } |x| \ge \alpha \end{cases}$$

(where $|x| = d(x, \mathbf{Z})$). Let $b_n = [\sin(n\pi\alpha)/(n\pi\alpha)]^2$ for $n \ne 0$, $b_0 = 1$; then

$$\int_0^1 \psi^2(x)\, dx = \sum_{n=-\infty}^{\infty} b_n{}^2 = \frac{2}{3\alpha} \qquad \text{and}$$

$$b_n{}^{-2} \le b_N{}^{-2} \le \left(\tfrac{\pi}{2}\right)^2 \qquad \text{for } n \le N.$$

Now choose any $N_r = 1, \ldots, N$ for $r = 1, \ldots, N$, and let

$$\tau_r(x) = \sum_{m=1}^{N_r} \frac{v_m}{b_m} e^{2\pi i m x} .$$

Then

$$s_r(x) = \sum_{m=1}^{N_r} v_m e^{2\pi i m x} = \psi * \tau_r(x) = \int_{-\alpha}^{\alpha} \psi(y)\, \tau_r(x-y)\, dy \ ,$$

so that by Hölder's inequality

$$|s_r(x)|^2 \le \int_{-\alpha}^{\alpha} \psi^2(y)\, dy \int_{-\alpha}^{\alpha} |\tau_r(x-y)|^2\, dy \ \le \ \frac{4N}{3} \int_{x-\alpha}^{x+\alpha} |\tau_r(y)|^2\, dy,$$

and hence,

$$|s_r(x_r)|^2 \le \frac{4N}{3} \int_{x_r-\alpha}^{x_r+\alpha} |\tau_r(y)|^2\, dy \ \le \frac{4N}{3} \int_{x_r-\alpha}^{x_r+\alpha} \sup_{r \le N} |\tau_r(y)|^2\, dy .$$

Therefore, using the fact that the intervals $(x_r - \alpha,\ x_r + \alpha)$ are disjoint and (1),

$$\sum_{r=1}^{N} |s_r(x_r)|^2 \leq \frac{4N}{3} \int_0^1 \sup_{r\leq N} |\tau_r(y)|^2 \, dy \leq \frac{4N}{3} C \sum_{m=1}^{N} \frac{v_m^{\,2}}{b_m^{\,2}}$$

$$\leq \frac{4N}{3} C(\tfrac{\pi}{2})^2 \sum_{m=1}^{N} v_m^{\,2} .$$

3. More maximal operators, weak (1, 1), Wiener-Wintner, and spectral continuity.

The maximal operators in (1)-(4) on the face of it are a rather diverse group, but because of Theorem 1 they are seen in fact to be extremely closely related. We now list some further variants, including a couple of ergodic-theoretic versions:

(5) Double Maximal Helical Transform on l^p. For $a \in l^p(\mathbf{Z})$ and $\theta \in \mathbf{R}$ define

$$H^{**}a(j) = \sup_{\substack{\theta\in\mathbf{R}\\ n>0}} \left| \sum_{k=-n}^{n}{}' \frac{e^{2\pi i(j-k)\theta} a_{j-k}}{k} \right| .$$

(6) Double maximal helical transform for a measure-preserving transformation. Let T: $X \to X$ be a measure-preserving transformation (m.p.t.) on a measure space $(X, \mathfrak{B}, \mu)$. For $f \in L^p(X, \mathfrak{B}, \mu)$ define

$$H^{**}f(x) = \sup_{n,\theta} \left| \sum_{k=-n}^{n}{}' \frac{e^{2\pi ik\theta} f(T^k x)}{k} \right| .$$

(7) Double maximal helical transform for a measure-preserving flow. Let $\{T_s: -\infty<s<\infty\}$ be a measure-preserving flow on a measure space $(X,\mathfrak{B},\mu)$. For $f\in L^p(X,\mathfrak{B},\mu)$ define

$$F^{**}f(x) = \sup_{\eta,\theta} \left| \int_{\eta\leq|s|\leq\frac{1}{\eta}} \frac{e^{2\pi is\theta}f(T_s x)}{s}\, ds \right| .$$

(8) Maximal Hilbert transform with a twist. For $f\in L^p[0, 1)$, define

$$D^*f(x) = \sup_{n\in\mathbf{Z}} \left| \int_0^1 \frac{e^{2\pi int}f(t)}{x-t}\, dt \right| .$$

THEOREM 2 [3, 2]: The operator in (4) is strong (p, p) for all p with $1<p<\infty$. The operator in (3) is not strong (p, p) for $1<p<2$. The strong type (p, p) property for each of the operators in (1), (4), (5), (6), (7), (8) is formally equivalent to strong type (p, p) for each of the others.

Proof: Strong (p, p) for (5) follows from that for (4) by a convolution trick from harmonic analysis. This then transfers to prove the estimate for (6), which in turn, using time δ maps in the flow, produces strong (p, p) for (7). Strong (p, p) for (8) for all $1<p<\infty$ was the main result of [14]; the equivalence of this with strong (p, p) for (1) is a direct consequence of classical calculations involving the Dirichlet kernel and Hilbert transform–see [13, p.102]. We sketch here a proof (given in [2] and avoiding use of the method of Hunt, Muckenhoupt, and Wheeden [15]) that strong (p, p) for (1) implies strong (p, p) for (4).

First notice that (8) being strong (p, p) implies that

$$\sup_{-1<\epsilon<1} \left| \int_{-\infty}^{\infty} \frac{e^{2\pi i\epsilon t} f(t)}{x-t}\, dt \right| \in L^p(\mathbf{R}) ,$$

with L^p norm bounded by a constant times the L^p norm of f. For making the change of variables $t=2^n s$ converts

$$\sup_{k\in\mathbf{Z}} \left| \int_{-2^n}^{2^n} \frac{e^{2\pi i \frac{k}{2^n} t} f(t)}{x-t}\, dt \right|$$

to two copies of the expression in (8), with L^p norm bounded by a constant (independent of n) times the L^p norm of f. (For $|x|>2^{2n}$, use the bound $C/|x|$; the region in between only enlarges the constant slightly.) Restricting k to be 2^r for some $r<n$ shows that the region of integration $[-2^n, 2^n]$ can be enlarged to $(-\infty, \infty)$. By monotone convergence we achieve the estimate when the supremum is taken over all dyadic rationals ϵ; the full estimate results from approximation, after for each fixed x restricting the t integration to a sufficiently large interval.

Given $a\in l^p(\mathbf{Z})$ define $f(x)=a_k$ for $k-1/8\leq x\leq k+1/8$, 0 otherwise. Then

$$C\|a\|_{l^p}^p \geq \int_{-\infty}^{\infty} \sup_{-1<\epsilon<1} \left| \int_{-\infty}^{\infty} \frac{e^{2\pi i\epsilon t} f(t)}{x-t}\, dt \right|^p dx$$

$$= \sum_{j=-\infty}^{\infty} \int_0^1 \sup_{-1<\epsilon<1} \left| \int_{-\infty}^{\infty} \frac{e^{2\pi i\epsilon t} f(t)}{x+j-t}\, dt \right|^p dx$$

$$= \sum_{j=-\infty}^{\infty} \int_0^1 \sup_{-1<\epsilon<1} \left| \sum_{k=-\infty}^{\infty} e^{2\pi i k\epsilon} a_k \int_{-\frac{1}{8}}^{\frac{1}{8}} \frac{e^{2\pi i\epsilon t}}{x+j-k-t}\, dt \right|^p dx$$

$$\geq \sum_{j=-\infty}^{\infty} \int_0^{\frac{1}{8}} \sup_{-1<\epsilon<1} \Bigg| \int_{-\frac{1}{8}}^{\frac{1}{8}} \Bigg\{ e^{2\pi i j\epsilon} a_j \frac{e^{2\pi i\epsilon t}}{x-t} +$$

$$\sum_{k=-\infty}^{\infty}{}' \frac{e^{2\pi i k\epsilon} a_k}{j-k} e^{2\pi i\epsilon t} \left[1 - \frac{x-t}{x+j-k-t} \right] \Bigg\} dt \Bigg|^p dx .$$

Because

$$\int_0^{\frac{1}{8}} \sup_{-1<\epsilon<1} \left| \int_{-\frac{1}{8}}^{\frac{1}{8}} \frac{e^{2\pi i\epsilon t}}{x-t}\, dt \right|^p dx$$

is a finite absolute constant by the opening remarks of the proof and

$$\int_{-\frac{1}{8}}^{\frac{1}{8}} e^{2\pi i\epsilon t}\, dt$$

is bounded below by an absolute constant, the result follows from noting that for $x\in[0,\,1/8]$

$$\sum_{j=-\infty}^{\infty} \sup_{-1<\epsilon<1} \left| {\sum_{k=-\infty}^{\infty}}' \frac{e^{2\pi ik\epsilon}a_k}{j-k} \left[\int_{-\frac{1}{8}}^{\frac{1}{8}} \frac{e^{2\pi i\epsilon t}(x-t)}{x+j-k-t}\, dt \right] \right|^p$$

$$\le C \sum_{j=-\infty}^{\infty} \sup_{-1<\epsilon<1} \left[{\sum_{k=-\infty}^{\infty}}' \frac{|a_k|}{|j-k|} \int_{-\frac{1}{8}}^{\frac{1}{8}} \frac{1}{|x+j-k-t|}\, dt \right]^p$$

$$\le C \sum_{j=-\infty}^{\infty} \left[{\sum_{k=-\infty}^{\infty}}' \frac{|a_k|}{(j-k)^2} \right]^p \le C \|a\|_{l^p}^p .$$

<u>REMARKS</u>: 1. Higher-dimensional versions of these inequalities can be achieved by beginning with the analogue of (3) and using a trick from [10] which breaks an integral over a square into the sum of integrals over two triangles.

2. In [2] some estimates are made on the constants involved, allowing one to show for example that if $f\in L\log^4 L$ then $\|H^{**}f\|_1<\infty$.

THEOREM 3 [3]: The double maximal helical transform $H^{**}f(x)$ in (6) does *not* satisfy a weak (1, 1) maximal estimate, that is, there is no constant C such that $\mu\{x: H^{**}f(x)>\lambda\}\leq\frac{C}{\lambda}\|f\|_1$ for all $f\in L^1(X,\mathfrak{B},\mu)$.

Proof: M. Lacey and M. Marcus (personal communication) nave provided an example of an independent identically distributed sequence $\{X_k\}$ for which weak (1, 1) fails. Using results of [17], they show that we can have $X_k\in L^1$ (in fact they are in L $\log\log^\beta$L for $0<\beta<1$) but with the associated double helical maximal function H^{**} equal to infinity almost everywhere. Also, Kolmogorov's example of a divergent Fourier series for an L^1 function may also be transferred to this setting. For if the operator H^{**} were weak (1, 1), then by arguments similar to ones used above (for example using time δ maps) we could conclude that the analogous operator F^{**} for continuous-parameter flows in (6) is also weak (1, 1), and in particular this would hold true for the translation flow on **R**. Because of the formula alluded to above, this would imply that the maximal operator for partial sums of Fourier series is also (1, 1), but this is not true.

THEOREM 4 [1]. The Wiener-Wintner property does *not* hold on L^1 (not even on L $\log\log^\beta$L for any $0<\beta<1$) for the helical transform: it is *not* true for each $f\in L^1$ that there is a single set of measure 0 in X off of which

$$H_\theta f(x) = \lim_{n\to\infty}\sum_{k=-n}^{n}{}' \frac{e^{ik\theta}f(T^kx)}{k}$$

exists for all θ.

Proof: The i.i.d. example of Lacey and Marcus is combined with a theorem of Billard [16, p. 58].

REMARK: On the other hand, it still seems likely that the Wiener-Wintner property does hold for all $f \in L^p$, $p>1$.

The helical transform can be used to compute the spectral measures of intervals. Combining formulas obtained by means of the functional calculus with the double maximal inequality for flows (strong (2, 2) for (7)), we can show that the spectral measure of a measure-preserving flow has a sort of continuity property at ∞. This strengthens, for this case, Gaposhkin's necessary and sufficient spectral condition for a flow of unitary operators to satisfy the Local Ergodic Theorem [12].

THEOREM 5 [3]: For a unitary flow $\{U_s: -\infty<s<\infty\}$ on $L^2(X,\mathfrak{B},\mu)$ with spectral measure E, $f \in L^2(X,\mathfrak{B},\mu)$, and any sequence $r_k \to \infty$, the properties (i) and (ii) below are equivalent; if $\{U_s\}$ comes from a measure-preserving flow on X, then both properties hold:

(i) $E(-r_k, 0)f(x) \to E(-\infty, 0)f(x)$ a.e. .

(ii) If $F_\theta f(x) = \lim_{n\to\infty} \int_{\frac{1}{n}\leq|s|\leq n} \frac{e^{2\pi i s\theta}\, U_s f(x)}{s}\, ds$ and $Ff = F_0 f$,

then

$$F_{r_k} f(x) \to Ff(x) + i\pi E\{0\}f(x) + 2i\pi E(-\infty, 0)f(x) \quad \text{a.e.} .$$

Proof. It follows from easy functional-calculus manipulations that

$$\frac{1}{i}[F_r - F] = \pi \int_{-\infty}^{\infty} [\operatorname{sgn}(\lambda + r) - \operatorname{sgn}\lambda]\, dE(\lambda)$$

$$= 2\pi E(-r, 0) + \pi(E\{0\} + E\{-r\}) .$$

From this equation the equivalence of statements (i) and (ii) for any unitary flow is clear. A dense set of functions satisfying (ii) is easily found, and then that both statements hold in the measure-preserving case follows from the maximal inequality for (7).

4. Lacunary sampling, kernel estimates, and quadratic variation.

In this section we investigate easier versions of the strong (2, 2) estimates for (1) and (3) above, namely for

$$S_0^* \hat{h}(x) = \sup_{r>0} \left| \sum_{j=-2^r}^{2^r} \hat{h}(j)\, e^{2\pi i j x} \right|$$

and

$$I_0^* h(j) = \sup_{r\geq 0} \left| \int_0^{2^{-r}} h(x)\, e^{-2\pi i j x}\, \mathrm{d}x \right| .$$

The fact that S_0^* is strong (2, 2) is a theorem of Kolmogorov. We show how strong (2, 2) for I_0^* follows from the same easy kernel estimates as used by Gaposhkin in [11]. Then Kolmogorov's theorem is a direct consequence, by the same path used in [3] to prove that (3) above implies (1) above. The same arguments apply to any lacunary sequences as well as 2^r and 2^{-r}; details may be seen in [19]. (The more common route to this theorem is via the Féjer-Lebesgue theorem giving the a.e. convergence of the Cesaro means of the Fourier series of L^1 functions, and the equivalence of the Cesaro method of summation with that of the 2^r'th partial sums for L^2 functions.)

THEOREM 6 [19]: There is a constant C such that

$$\|\Gamma_0^* h(j)\|_{l^2}^2 \leq C\|h\|_{L^2}^2 \quad \text{for all } h\in L^2[0, 1) .$$

Proof. Let

$$K_n(p) = \frac{1}{2n+1}\sum_{m=-n}^{n} \delta_{\{m\}}(p), \qquad \text{so that}$$

$$\hat{K}_n(t) = \frac{1}{2n+1}\sum_{m=-n}^{n} e^{2\pi imt} .$$

Then

$$\int_0^{2^{-r}} h(t)\, e^{-2\pi ijt} dt = \int_0^{2^{-r}} (1-\hat{K}_{2^r}(t))\, h(t)\, e^{-2\pi ijt} dt$$

$$- \int_{2^{-r}}^{1} \hat{K}_{2^r}(t)\, h(t)\, e^{-2\pi ijt} dt \; + \; \int_0^1 \hat{K}_{2^r}(t)\, h(t)\, e^{-2\pi ijt} dt.$$

We also have the two easy estimates

(i) $|1-\hat{K}_n(t)| \leq \frac{1}{2n+1}\sum_{m=-n}^{n} |\, 1 - e^{2\pi imt}| \leq \frac{1}{2n+1} n(n+1)\, C2\pi t \leq Cnt$ and

(ii) $|\hat{K}_n(t)| = \frac{1}{2n+1} |(1 - e^{2\pi i(2n+1)t})\,(1 - e^{-2\pi it})^{-1}|$

$$\leq \frac{1}{2n+1}\, (2)\, \frac{1}{Ct} \leq C(nt)^{-1} .$$

Since $(a+b+c)^2 \leq 8(a^2+b^2+c^2)$, it is enough to prove a strong (2, 2) estimate for the suprema of each of the three terms on the right.

To bound the first of these three terms, note that

$$\left\| \sup_r \left| \int_0^{2^{-r}} (1 - \hat{K}_{2^r}(t))\, h(t)\, e^{-2\pi i j t} dt \right| \right\|_{l^2}^2$$

$$\leq \sum_{j=-\infty}^{\infty} \sum_{r=1}^{\infty} \left| \int_0^{2^{-r}} (1 - \hat{K}_{2^r}(t))\, h(t)\, e^{-2\pi i j t} dt \right|^2$$

$$= \sum_{r=1}^{\infty} \left\| \int_0^{2^{-r}} (1 - \hat{K}_{2^r}(t))\, h(t)\, e^{-2\pi i j t} dt \right\|_{l^2}^2$$

$$= \sum_{r=1}^{\infty} \left\| [(1 - \hat{K}_{2^r})\, h\, \chi_{[0,2^{-r}]}]\hat{} \right\|_{L^2}^2$$

$$= \sum_{r=1}^{\infty} \int_0^1 \left| (1 - \hat{K}_{2^r}(t))\, h(t)\, \chi_{[0,2^{-r}]}(t)\, dt \right|^2$$

$$= \int_0^1 |h(t)|^2 \sum_{r=1}^{\infty} \left| 1 - \hat{K}_{2^r}(t) \right|^2 \chi_{[0,2^{-r}]}(t)\, dt$$

$$\leq \|h\|_{L^2}^2 \left\| \sum_{r=1}^{\infty} (1 - \hat{K}_{2^r})\, \chi_{[0,2^{-r}]} \right\|_{\infty}^2$$

$$\leq \|h\|_{L^2}^2 \left\| \sum_{r=1}^{\infty} C 2^r\, t\, \chi_{[0,2^{-r}]}(t) \right\|_{\infty}^2 .$$

If $2^{-s} \leq t \leq 2^{-s+1}$, then this is bounded by

$$\|h\|_{L^2}^2 \left\| \sum_{r=1}^{s-1} C 2^r\, t \right\|_{\infty}^2 \leq \|h\|_{L^2}^2\, 2\, C\,, \quad \text{independently of } s.$$

A similar argument, except with the estimate (i) replaced by (ii), bounds the second term. To deal with the third term, we use the strong type (2,2) Hardy-Littlewood maximal inequality on l^2.

$$\left\| \sup_r \left| \int_0^1 \hat{K}_{2^r}(t)\, h(t)\, e^{-2\pi i j t} dt \right| \right\|_{l^2}^2 \leq \left\| \sup_n \left| \int_0^1 \hat{K}_n(t)\, h(t)\, e^{-2\pi i j t} dt \right| \right\|_{l^2}^2$$

$$= \left\| \sup_n |[K_n * \hat{h}](j)| \right\|_{l^2}^2 = \left\| \sup_n \left| \frac{1}{2n+1} \sum_{m=-n}^{n} \hat{h}(j+m) \right| \right\|_{l_2}^2$$

$$\leq C \, ||\hat{h}||_{l^2}^2 = C \, ||h||_{L^2}^2 \ .$$

THEOREM 7 (Kolmogorov): There is a constant C such that

$$\left| S_0^* h \right|_{L^2}^2 \leq C ||h||_{L^2}^2 \text{ for all } h \in L^2[0, 1) \ .$$

Proof. We again pass through the discrete equivalent (2). Suppose first that h is a nonnegative step function, taking the value v_m on the interval $[(m-1)/2^N, m/2^N)$, $m=1,\ldots,2^N$. Then

$$I_0^* h(j) = \sup_{N \geq r \geq 0} \left| \int_0^{2^{-r}} h(x) \, e^{-2\pi i j x} \, dx \right|$$

$$= \sup_{N \geq r \geq 0} \left| \sum_{m=1}^{2^{N-r}} v_m \int_{\frac{m-1}{2^N}}^{\frac{m}{2^N}} e^{-2\pi i j x} \, dx \right|$$

$$= \frac{1}{2\pi |j|} \left| \sum_{m=1}^{2^{N_j}} v_m \, e^{-2\pi i j (m-1)/2^N} (e^{-2\pi i j/2^N} - 1) \right|$$

$$= \frac{1}{2\pi |j|} \left| \sum_{m=1}^{2^{N_j}} v_m \, \omega_r{}^{m-1} (\omega_r - 1) \right| ,$$

where $j = k \cdot 2^N + s$, $\omega_s = e^{2\pi i s/2^N}$, and N_j is chosen so as to achieve the supremum. The maximizing choice of N_j depends only on the congruence class of j mod 2^N; thus we replace 2^{N_j} by 2^{N_s}. Therefore

$$\frac{1}{2^N} \sum_{m=1}^{2^N} |v_m|^2 \geq C \left| I_0^* h \right|_{l^2}^2$$

$$= \sum_{k=-\infty}^{\infty} \sum_{s=1}^{2^N} \frac{1}{(k2^N + s)^2} \left| \sum_{m=1}^{2^{N_s}} v_m \omega_s{}^{m-1} (\omega_s - 1) \right|^2$$

$$\geq C \sum_{s=1}^{2^N} \frac{1}{sr^2} \left| (\omega_s - 1) \sum_{m=1}^{2^{N_s}} v_m \omega_s^{\,m-1} \right|^2$$

$$\geq \frac{C}{2^{2N}} \sum_{s=1}^{2^N} \left| \sum_{m=1}^{2^{N_s}} v_m \omega_s^{\,m-1} \right|^2 ,$$

since we may just as well sum over s in the range $-2^{N-1} \leq s < 2^{N-1}$, where $|\omega_s - 1|/s \geq C/2^N$.

Now in order to prove Kolmogorov's Theorem, given $\hat{h} \in l^2(\mathbf{Z})$, $K \geq 1$, and $x \in [0, 1)$, let

$$T_K^* \hat{h}(x) = \sup_{1 \leq r \leq K} \left| \sum_{j=-2^r}^{2^r} \hat{h}(j)\, e^{2\pi i j x} \right| = \left| \sum_{j=-2^{r(x)}}^{2^{r(x)}} \hat{h}(j)\, e^{2\pi i j x} \right| .$$

Since this is a continuous function of x, we may choose $N \geq K$ large enough that

$$\int_0^1 \left| T_K^* \hat{h}(x) \right|^2 dx$$

is arbitrarily well approximated by a Riemann sum

$$\sum_{s=1}^{2^N} \left| \sum_{j=-2^{N_s}}^{2^{N_s}} \hat{h}(j) \omega_s^{\,j} \right|^2 \cdot \frac{1}{2^N} .$$

Applying the result of the first paragraph of the proof twice, to the negative and nonnegative ranges of j, this is bounded by

$$C \sum_{j=-2^N}^{2^N} \left| \hat{h}(j) \right|^2 \leq C \| h \|_2^2 .$$

By means of these same kernel estimates, we can strengthen the Hardy-Littlewood strong (2, 2) inequality on l^2 to a quadratic-variation statement of the kind proved by Bourgain for averages along sequences such as the squares or the primes. A similar quadratic-variation strengthening of the Pointwise Ergodic Theorem is an immediate corollary. The basic idea behind these statements is in [4]; the following might serve as an exposition of part of the method of [4], by showing how it works in a particularly simple case. In particular, in this situation we obtain in the inequality simply a constant factor rather than an expression that tends to zero when divided by K.

For a function ϕ on $\mathbf{Z}$, define

$$A_n\phi(j) = \frac{1}{2n+1}\sum_{m=-n}^{n}\phi(p) = (K_n*\phi)(j) .$$

We use the fact that A_n is strong (2, 2) to prove the following stronger statement. (That it in fact is stromger may be seen by fixing $n_0=1$ and fixing n_1 at an arbirary large value.)

<u>THEOREM</u> <u>8</u>·<u>[19]</u>: There is a constant C such that whenever $\{n_k\}$ is a sequence of positive integers satisfying $n_{k+1}>n_k^8$ for all k, then for all $K>0$ and all $\phi\in l^2$

$$(*) \qquad \sum_{k=1}^{K}\left\|\max_{n_{k-1}\le n<n_k}\left|A_n\phi - A_{n_{k-1}}\phi\right|\right\|_{l^2}^2 \le C\,\|\phi\|_{l^2}^2 .$$

Proof. In a low-grade version of Bourgain's approach via Fourier analysis and the circle method, we take the Fourier transform to transfer the problem to [0,1), break the interval into pieces, and solve the problem on each piece.

Since

$$A_n\phi(j) = \int_0^1 \hat{K}_n(t)\ \hat{\phi}(t)\ e^{2\pi ijt}dt,$$

the left-hand side of (*) can be written as

$$(**)\quad \sum_{k=1}^{K}\left\| \max_{n_{k-1}\le n<n_k}\left|\int_0^1 (\hat{K}_n - \hat{K}_{n_{k-1}})(t)\,\hat{\phi}(t)\,e^{2\pi ijt}dt\right|\right\|_{l^2}^2 .$$

To bound this expression we will break the interval [0,1) into three pieces defined by

S_k = characteristic function of $[0,\ n_k^{-2})$

M_k = characteristic function of $[n_k^{-2},\ n_{k-1}^{-1/4})$

L_k = characteristic function of $[n_{k-1}^{-1/4},\ 1)$.

It is clear that, because we have chosen $\{n_k\}$ so that $n_{k+1}>n_k^8$, for each $t\in[0,\ 1)$ M_k is non-zero for at most two values of k. The point of this division will become clear when we show that, for $n_{k-1}\le n<n_k$, $\hat{K}_n$ and $\hat{K}_{n_{k-1}}$ are small where $L_k\neq 0$, so that (**) can be bounded easily there. Furthermore, where $S_k\neq 0$, $\hat{K}_n$ and $\hat{K}_{n_{k-1}}$ are both close to 1, and therefore $\hat{K}_n - \hat{K}_{n_{k-1}}$ is small, so that (**) may also be bounded easily on that interval. Where $M_k\neq 0$ there is no helpful bound on the kernels, but we may use the discrete version of the strong (2,2) maximal inequality of Hardy and Littlewood. At this point it will be crucial that the intervals determined by $M_k\neq 0$ do not overlap too much.

To carry out the above plan, we write (**) as

$$\sum_{k=1}^{K}\left\| \max_{n_{k-1}\le n<n_k}\left|\int_0^1 (\hat{K}_n - \hat{K}_{n_{k-1}})(t)\ \hat{\phi}(t)\ (S_k + M_k + L_k)(t)\ e^{2\pi ijt}dt\right|\right\|_{l^2}^2$$

$$\leq C\sum_{k=1}^{K}\left\| \max_{n_{k-1}\leq n<n_k}\left|\int_0^1(\hat{K}_n - \hat{K}_{n_{k-1}})(t)\ \hat{\phi}(t)\ S_k(t)\ e^{2\pi ijt}dt\right|\ \right\|_{l^2}^2$$

$$+\ C\sum_{k=1}^{K}\left\| \max_{n_{k-1}\leq n<n_k}\left|\int_0^1(\hat{K}_n - \hat{K}_{n_{k-1}})(t)\ \hat{\phi}(t)\ M_k(t)\ e^{2\pi ijt}dt\right|\ \right\|_{l^2}^2$$

$$+\ C\sum_{k=1}^{K}\left\| \max_{n_{k-1}\leq n<n_k}\left|\int_0^1(\hat{K}_n - \hat{K}_{n_{k-1}})(t)\ \hat{\phi}(t)\ L_k(t)\ e^{2\pi ijt}dt\right|\ \right\|_{l^2}^2 .$$

Ignoring the constant factors, the first of these terms may be estimated by

$$\sum_{k=1}^{K}\sum_{j=-\infty}^{\infty}\sum_{n=n_{k-1}}^{n_k-1}\left|\int_0^1(\hat{K}_n - \hat{K}_{n_{k-1}})(t)\ \hat{\phi}(t)\ S_k(\mathrm{t})\ \mathrm{e}^{2\pi ijt}dt\right|^2$$

$$=\sum_{k=1}^{K}\sum_{n=n_{k-1}}^{n_k-1}\left\|\int_0^1(\hat{K}_n - \hat{K}_{n_{k-1}})(t)\ \hat{\phi}(t)\ S_k(\mathrm{t})\ \mathrm{e}^{2\pi ijt}dt\right\|_{l^2}^2$$

$$=\sum_{k=1}^{K}\sum_{n=n_{k-1}}^{n_k-1}\left\|[(\hat{K}_n - \hat{K}_{n_{k-1}})\hat{\phi}\,S_k]^{\check{}}\right\|_{l^2}^2$$

$$=\sum_{k=1}^{K}\sum_{n=n_{k-1}}^{n_k-1}\left\|(\hat{K}_n - \hat{K}_{n_{k-1}})\hat{\phi}\,S_k\right\|_{L^2}^2$$

$$\leq \|\hat{\phi}\|_{L^2}^2\sum_{k=1}^{K}\sum_{n=n_{k-1}}^{n_k-1}\left\|(\hat{K}_n - \hat{K}_{n_{k-1}})\,S_k\right\|_{\infty}^2$$

$$= \|\phi\|_{l^2}^2\sum_{k=1}^{K}\sum_{n=n_{k-1}}^{n_k-1}\left\|(\hat{K}_n - \hat{K}_{n_{k-1}})\,S_k\right\|_{\infty}^2 .$$

Because of the kernel estimate (i) above, since $n_{k-1}\leq n<n_k$, we have $\left|\hat{K}_n(t) - \hat{K}_{n_{k-1}}(t)\right| \leq \left|1 - \hat{K}_n(t)\right| + \left|1 - \hat{K}_{n_{k-1}}(t)\right| \leq Cnt$, and hence $\left\|(\hat{K}_n - \hat{K}_{n_{k-1}})\,S_k\right\|_{\infty}^2 \leq [Cn(n_k^{-2})]^2 \leq [Cn(n^{-2})]^2 = C^2n^{-2}$. It follows that this term is bounded by

$$\|\phi\|_{l^2}^2\ C^2\sum_{k=1}^{K}\sum_{n=n_{k-1}}^{n_k-1} n^{-2} \leq C\,\|\phi\|_{l^2}^2 .$$

Again dropping constant factors and applying Hardy-Littlewood at the appropriate point, the second term is estimated by

$$\sum_{k=1}^{K}\left|\max_{n_{k-1}\le n<n_k}\left|\int_0^1 \hat{K}_n(t)\,\hat{\phi}(t)\,M_k(t)\,e^{2\pi ijt}\mathrm{d}t\right|\right|_{l^2}^2$$

$$\le \sum_{k=1}^{K}\left|\sup_n\left|\int_0^1 \hat{K}_n(t)\,\hat{\phi}(t)\,M_k(t)\,e^{2\pi ijt}\mathrm{d}t\right|\right|_{l^2}^2$$

$$= \sum_{k=1}^{K}\left|\sup_n\left|K_n * [\hat{\phi}M_k]^{\check{}}\right|\right|_{l^2}^2 \le C\sum_{k=1}^{K}\left|[\hat{\phi}M_k]^{\check{}}\right|_{l^2}^2$$

$$= C\sum_{k=1}^{K}\left|\hat{\phi}M_k\right|_{L^2}^2 = C\int_0^1\left|\hat{\phi}(t)\right|^2\sum_{k=1}^{K}M_k(t)\,\mathrm{d}t$$

$$\le C\left|\hat{\phi}\right|_{L^2}^2 = \left|\phi\right|_{l^2}^2 .$$

The third term, up to constant factors, is bounded by

$$\sum_{k=1}^{K}\left|\max_{n_{k-1}\le n<n_k}\left|\int_0^1 \hat{K}_n(t)\,\hat{\phi}(t)\,L_k(t)\,e^{2\pi ijt}\mathrm{d}t\right|\right|_{l^2}^2$$

$$\le \sum_{k=1}^{K}\sum_{n=n_{k-1}}^{n_k-1}\left|\int_0^1 \hat{K}_n(t)\,\hat{\phi}(t)\,L_k(t)\,e^{2\pi ijt}\mathrm{d}t\right|_{l^2}^2$$

$$= \sum_{k=1}^{K}\sum_{n=n_{k-1}}^{n_k-1}\left|[\hat{K}_n\,\hat{\phi}\,L_k]^{\check{}}\right|_{l^2}^2 = \sum_{k=1}^{K}\sum_{n=n_{k-1}}^{n_k-1}\left|\hat{K}_n\hat{\phi}\,L_k\right|_{L^2}^2$$

$$\le \left|\hat{\phi}\right|_{L^2}^2\sum_{k=1}^{K}\sum_{n=n_{k-1}}^{n_k-1}\left|\hat{K}_n\,L_k\right|_{\infty}^2 .$$

Using the kernel estimate (ii) from above

$$\left|\hat{K}_n\,L_k\right|_{\infty}^2 < C^2(n\,n_{k-1}^{-1/4})^{-2} = C^2\frac{\sqrt{n_{k-1}}}{n^2} \le C^2\frac{\sqrt{n}}{n^2} = C^2 n^{-3/2}$$

(since $n_{k-1}\le n$), and hence the third term is bounded by a universal constant times

$$\sum_{k=1}^{K} \sum_{n=n_{k-1}}^{n_k-1} C^2 n^{-3/2} \|\phi\|_{l^2}^2 \leq C\|\phi\|_{l^2}^2 .$$

For a measure-preserving transformation $T:X\to X$ and measurable function f on X, let $B_n f(x)=\frac{1}{n}\sum_{k=0}^{n-1} f(T^k x)$ denote the n'th Cesaro average of f along the orbit of T. The following theorem follows readily from the previous one by the usual transference method, and the ordinary Pointwise Ergodic Theorem on L^2 is an immediate consequence.

THEOREM 7: There is a constant C such that whenever $\{n_k\}$ is a sequence of positive integers satisfying $n_{k+1}>n_k^8$ for all k, then for all $K>0$ and all $f\in L^2(X,\mathfrak{B},\mu)$

$$\sum_{k=1}^{K} \left\| \max_{n_{k-1}\leq n<n_k} |B_n f - B_{n_{k-1}} f| \right\|_{L^2}^2 \leq C\|f\|_{L^2}^2 .$$

REFERENCES

1. I. Assani, The Wiener-Wintner property for the helical transform of the shift on $[0, 1]^{\mathbb{Z}}$, preprint.

2. I. Assani, The helical transform and the a.e. convergence of Fourier series, preprint.

3. I. Assani and K. Petersen, The helical transform as a connection between ergodic theory and harmonic analysis, to appear in Trans. Amer. Math. Soc.

4. J. Bourgain, On the maximal ergodic theorem for certain subsets of the integers, Israel J. Math. 61 (1988), 39-72.

5. J. Campbell and K. Petersen, The spectral measure and Hilbert transform of a measure-preserving transformation, Trans. Amer. Math. Soc. 313 (1989), 121-129.

6. A. P. Calderón, Ergodic theory and translation invariant operators, Proc. Nat. Acad. Sci. U.S.A. 59 (1968), 349-353.

7. L. Carleson, On convergence and growth of partial sums of Fourier series, Acta Math. 116 (1966), 135-157.

8. M. Cotlar, A unified theory of Hilbert transforms and ergodic theorems, Rev. Mat. Cuyana 1 (1955), 105-167.

9. H. Davenport and H. Halberstam, The values of a trigonometrical polynomial at well spaced points, Mathematika 13 (1966), 91-96.

10. C. Fefferman, On the convergence of multiple Fourier series, Bull. Amer. Math. Soc. 77 (1971), 744-745.

11. V.F. Gaposhkin, Individual ergodic theorem for normal operators in L^2, Functional Anal. Appls. 15 (1981), 18-22.

12. V. F. Gaposhkin, The Local Ergodic Theorem for groups of unitary operators and second order stationary processes, Math. USSR Sbornik 39 (1981), 227-242.

13. A. M. Garsia, *Topics in Almost Everywhere Convergence*, Markham Pub. Co., Chicago, 1970.

14. R. Hunt, On the convergence of Fourier series, *Orthogonal Expansions and their Continuous Analogues*, D.T. Haimo, ed., S. Ill. Univ. Press, Carbondale, 1968, 235-255.

15. R. Hunt, B. Muckenhoupt, and R. Wheeden, Weighted norm inequalities for the conjugate function and Hilbert transform, Trans. Amer. Math. Soc. 176 (1973), 227-251.

16. J.P. Kahane, *Some Random Series of Functions*, Cambridge Univ. Press, Cambridge, 1985.

17. M.B. Marcus, *ξ-radial Processes and Random Fourier Series*, Memoirs Amer. Math. Soc. 181, Providence, R.I., 1987.

18. K. Petersen, Another proof of the existence of the ergodic Hilbert transform, Proc. Amer. Math. Soc. 88 (1983), 39-43.

19. H. White, The Pointwise Ergodic Theorem and related analytic inequalities, Master's Thesis, U.N.C., 1989.

Department of Mathematics
CB#3250, Phillips Hall
University of North Carolina
Chapel Hill, NC 27599 USA

On Hopf's Ergodic Theorem for Particles with Different Velocities

by

Alexandra Bellow[1] and Ulrich Krengel[2]

Summary

In 1936 E. Hopf [H2] answered a limiting problem on the distribution of a cloud of particles in a box with the help of a general L_1-mean ergodic theorem. We answer his question whether almost everywhere convergence holds in his setting in the negative. The corresponding local ergodic theorem is also studied.

1. Hopf's Billiard

In the thirties E. Hopf [H1], [H2] studied the "method of arbitrary functions" initiated by Poincaré, Borel and others. The aim was to explain why the probabilities of certain random events were frequently essentially independent of the initial conditions of an experiment. One of his key examples in [H2] was a model describing the motion of many particles in a box with elastic reflection at the sides of the box. He showed that, if the distribution of the speed of the particles is given by a density, the distribution of the particles in the box converges weakly (without averaging).

Formally, Hopf considered an abstract finite measure space $(\Omega,\mathfrak{F},\mu)$ and a measureable measure preserving flow $\{\tau_t, -\infty < t < \infty\}$. If $\omega \in \Omega$ represents the location and direction of a particle moving at unit speed, $\tau_t\omega$ is the vector giving its location and direction at time t. If v is the velocity of the particle, its position in Ω at time t is $\tau_{tv}\omega$.

Let $\tilde{\Omega} = \Omega \times [0,\infty)$, and let $\tilde{\mu} = \mu \otimes \lambda$ be the product of μ and the Lebesgue measure λ. If a probability density f in $\tilde{\Omega}$ describes the distribution of a cloud of particles at time 0 and if H is a measureable subset of $\tilde{\Omega}$, the relative proportion of the particles in H at time t is given by

$$\int\int f(\omega,v)1_H(\tau_{tv}\omega,v)\mu(d\omega)\lambda(dv) = \int\int f(\tau_{-tv}\omega,v)1_H(\omega,v)d\mu d\lambda.$$

The convergence of this quantity for $t \to \infty$ follows by applying the following theorem of Hopf to the flow with reversed time.

---

[1] The visit of this author at the University of Göttingen was supported by a Humbolt prize. This research was also partially supported by NSF grant DMS–8910947.

[2] This work was completed during a visit of the second author at the School of Mathematics of the Georgia Institute of Technolgoy.

ISBN 0-12-085520-8

Theorem 1 (Hopf, 1936): *Let $\{\tau_t,\ 0 \leq t < \infty\}$ be a measurable measure preserving flow in a finite measure space $(\Omega,\mathcal{F},\mu)$. For $f \in L_1(\tilde{\mu})$ and $h \in L_\infty(\tilde{\mu})$ put*

$$T_t f(\omega) = \int_0^\infty f(\tau_{tv}\omega,v)h(\omega,v)\lambda(dv).$$

Then $T_t f$ converges in L_1-norm for $t \to \infty$.

The idea of the proof is to consider functions of the form $f(\omega,v) = g(\omega)1_{[a,b]}(v)$ and $h(\omega,v) = d(\omega)1_{[c,d]}(v)$ first. If $[\alpha,\beta] = [a,b] \cap [c,d]$, a substitution gives

$$T_t f(\omega) = d(\omega)\cdot\frac{1}{t}\int_{\alpha t}^{\beta t} g(\tau_s\omega)ds,$$

and the usual ergodic theorem yields convergence in L_1. For general f,h, use approximations by linear combinations.

Hopf conjectured that $T_t f$ converges a.e. for $f \in L_1(\tilde{\mu})$, $h \in L_\infty(\tilde{\mu})$. The main purpose of this paper is to provide a counterexample to this conjecture.

2. A Counterexample

We shall show that $T_t f$ need not converge a.e. even if f is an indicator function of a set of finite $\tilde{\mu}$-measure (and hence belongs to all spaces $L_p(\tilde{\mu})$ $(1 \leq p \leq \infty)$). Moreover, we shall have $h \equiv 1$ in our example.

Let $\Omega = [0,1)$, and let μ be the Lebesgue measure on Ω. The flow is given by $\tau_t\omega = \omega + t$ (with addition mod 1). The indicator function $f = 1_A$ will be constructed in a sequence of steps.

Step 1: We construct a small set A_1 and integers $n_1,\ldots,n_m$ such that $\sup_{i\leq m} T_{n_i}1_{A_1}$ is large on Ω.

Start with a large integer $K = K_1$. Define

$$\epsilon = \epsilon_1 = 1/K, \quad \beta = \beta_1 = 1/K^2.$$

Let $\alpha > 0$ be an irrational number. Divide the interval $[0,1)$ into $m = 2/\beta = 2K^2$ disjoint subintervals $[a_k,b_k)$, $k = 1,\ldots,m$ putting

$$a_k = (k-1)\beta/2, \quad b_k = k\beta/2.$$

As the sequence $\{n\alpha\}$ is dense in $[0,1)$ (mod 1) we can inductively find integers $K \le n_1 < n_2 < \cdots < n_m$ such that

$$(1) \qquad n_k\alpha \in (-a_k, -a_k + \beta/4) \quad (\text{mod } 1).$$

Call $N = n_m$. Choose now $0 < \gamma < 1$ with

$$(2) \qquad \gamma < \beta/(4N),$$

and let

$$(3) \qquad c = [K/\gamma]$$

be the integral part of K/γ. Note that since $K/\gamma - 1 < c \le K/\gamma$ we have

$$(4) \qquad K-1 < K-\gamma < c\gamma \le K$$

and hence

$$(5) \qquad \beta c\gamma \le \frac{1}{K^2} K = \frac{1}{K} = \epsilon.$$

Let now $p \in \mathbb{N}$ and

$$F = \bigcup_{i=p}^{p+c-1} [i+\alpha,\ i+\alpha+\gamma).$$

Define $A_1 = [0,\beta) \times F$ and $f_1 = 1_{A_1}$. Note that

$$||f_1||_1 = \tilde{\mu}(A_1) = \beta c\gamma \le \epsilon.$$

For each $\omega \in \Omega$ there is an integer k with $1 \le k \le m$ and $\omega \in [a_k, b_k)$. For each $v \in F$ there is some integer $i \in [p,\ p+c-1]$ such that

$$i + \alpha \leq v < i + \alpha + \gamma.$$

We have (mod 1)

$$\begin{aligned}\tau_{n_k v}\omega &= n_k v + \omega = n_k(i + \alpha) + n_k(v - (i+\alpha)) + a_k + (\omega - a_k) \\ &= n_k\alpha + a_k + n_k(v - (i+\alpha)) + (\omega - a_k) \\ &< \beta/4 + N\gamma + \beta/2 < \beta/4 + \beta/4 + \beta/2 = \beta.\end{aligned}$$

Hence, for $\omega \in [a_k, b_k)$ we have

$$\tau_{n_k v}(\omega) \in [0,\beta) \quad \text{for all } v \in F.$$

It follows that

$$f_1(\tau_{n_k v}\omega, v) = 1 \quad \text{for all } v \in F.$$

Hence

$$T_{n_k} f_1(\omega) = \int_F f_1(\tau_{n_k v}\omega, v)\lambda(dv) = \lambda(F) = c\gamma > K-1.$$

As $\omega \in \Omega$ was arbitrary,

$$\sup_{k \leq m} T_{n_k} f_1 \geq K-1 \quad \text{on } \Omega.$$

In steps 2,3,... the same method is used to construct sets $A_2, A_3, \ldots$. In step j one starts with a large number K_j and obtains A_j with $\tilde{\mu}(A_j) \leq 1/K_j$. If the number $p = p_j$ used in step j is larger than the number $p+c-1$ from the previous step, the sets $A_1, A_2, \ldots$ are disjoint. If the K_j's grow fast enough the union A of the A_j's will be arbitrarily small and

$$\limsup_{n \to \infty} T_n 1_A = \infty.$$

As $T_n 1_A$ converges in L_1-norm, this contradicts almost everywhere convergence.

Remarks: 1) The same construction allows to show in fact the following: Given any increasing sequence of positive integers $\{n_k\}$, and $\epsilon > 0$, there exists an indicator function $f = 1_A$ with $\tilde{\mu}(A) < \epsilon$ and $\limsup T_{n_k} f = +\infty$ a.e. (all we need is an irrational $\alpha > 0$ such that $[n_k \alpha]$ is uniformly distributed mod 1).

2) The construction can be slightly simplified, if one wants to obtain only $\limsup_{t \to \infty} = \infty$ a.e. with t ranging through *all* reals.

3) It is possible to construct an (unbounded) integrable f with $\limsup T_t f = \infty$ having the additional property that $f(\omega, v) = 0$ for all $\omega \in \Omega$ and $v \geq 1$. (Actually, this is simpler.)

3. The Local Ergodic Theorem

The following is a local version of Hopf's above theorem.

<u>Theorem 2</u>: *Under the assumptions of Theorem 1, $T_t f$ converges in $L_1(\mu)$-norm to*

$$\int_0^\infty f(\cdot, v) h(\cdot, v) dv$$

for $t \to 0+0$.

The proof uses the same approximation and the norm-continuity of $g \circ \tau_t$. In fact, this argument also shows that if

$$f_t(\omega, v) = f(\tau_{tv}\omega, v),$$

then f_t converges to f in $L_1(\tilde{\mu})$-norm.

We now show by example that $T_t f$ need not converge a.e. for $t \to 0+0$:

Ω, μ and τ_t are as in Section 2. We first show that for any $\epsilon > 0$ there exist an $f_\epsilon \in L_1^+(\tilde{\mu})$ with

$$\|f_\epsilon\|_1 = \int f_\epsilon d\tilde{\mu} < \epsilon$$

and positive numbers $t_1, \ldots, t_m \leq \epsilon$ such that

(6) $$\sup\left\{T_{t_1}f_\epsilon,\ T_{t_2}f_\epsilon,\ldots,T_{t_m}f_\epsilon\right\} \geq \frac{1}{2\epsilon}\ .$$

We may assume that $K = 1/\epsilon$ is a large integer. Choose

$$c = K^2,\quad \beta = \frac{1}{K^2},\quad \gamma = \frac{1}{2K},\quad t_1 = \epsilon = \frac{1}{K}.$$

Set $$g_\epsilon(\omega) = c1_{[0,\beta]}(\omega)$$

$$f_\epsilon(\omega,v) = g_\epsilon(\omega)1_{[K,K+\gamma]}(v).$$

Then $$\|f_\epsilon\|_1 = c\beta\gamma = 1/(2K) < \epsilon.$$

It follows from $t_1K = 1 = 0 \pmod 1$ and $t_1\gamma = 1/(2K^2) = \beta/2$ that t_1v belongs (mod 1) to $[0,\beta/2]$ for $v \in [K,K+\gamma]$. Hence $\tau_{t_1v}\omega$ belongs to $[0,\beta]$ for $K \leq v \leq K+\gamma$ and $\omega \in [0,\beta/2]$. This implies $T_{t_1}f_\epsilon(\omega) = c\gamma = K^2/(2K) = K/2$ for $\omega \in [0,\beta/2]$.

As in Section 2, $\Omega = [0,1)$ can be written as the disjoint union of $m = 2K^2$ intervals $[a_\nu,b_\nu) = [(\nu-1)\beta/2,\ \nu\beta/2)$, $\nu = 1,\ldots,m$. Determine $t_\nu \leq t_1$ with $(t_1-t_\nu)K = (\nu-1)\beta/2 = a_\nu$. For $\omega = a_\nu$ we have (mod 1) $\tau_{t_\nu K}\omega = \omega + t_\nu K = a_\nu - (t_1-t_\nu)K = 0$. Since $t_\nu\gamma \leq t_1\gamma \leq \beta/2$, it follows that for $\omega \in [a_\nu,b_\nu)$ and $K \leq v \leq K+\gamma$

$$T_{t_\nu v}\omega \in [0,\beta].$$

Hence, $T_{t_\nu}f_\epsilon(\omega) \geq K/2$ for $\omega \in [a_\nu,b_\nu)$. This proves (6).

If $\epsilon_1,\epsilon_2,\ldots$ is a sequence of positive numbers with convergent sum,

$$f = \sum_{i=1}^{\infty} f_{\epsilon_i}$$

belongs to $L_1^+(\tilde{\mu})$ and $\limsup_{t\to 0} T_tf = \infty$. This shows that T_tf does not converge a.e.

Remark: One can show that there even exists an indicator function $f = 1_B$ with $\tilde{\mu}(B)$ arbitrarily small and $\limsup_{t\to 0} T_tf = \infty$ a.e. We just sketch the idea: For large $K = K_1$, put $c = K^2$, $\beta = 1/K^2$, $\gamma = 1/(4K)$, $k = 4K^4$, $\ell = c + k - 1$. Let $f_1(\omega,v) = 1_{[0,\beta]}(\omega)1_F(v)$ where

$$F_1 = F = \bigcup_{j=k}^{\ell} [jK, jK+\gamma).$$

Then $||f_1||_1 < 1/K$. Using arguments similar to those above it can be seen that there exist $t_1 = 1/K > t_2 > \cdots t_m > 0$ with

$$\sup\{T_{t_1}f_1,\ldots,T_{t_m}f_1\} \geq K/4 \quad \text{on } \Omega.$$

Repeat this with a larger $K = K_2 > \ell$ to find f_2, etc. Let $B_i = \text{supp}(f_i)$ and $B = \cup B_i$.

References

[H1] E. Hopf: On causality, statistics and probability. J. of Math. and Phys. 13, (1934), 51-102.

[H2] E. Hopf: Ueber die Bedeutung der willkuerlichen Funktionen fuer die Wahrscheinlichkeitstheorie. Jahresber. Dtsch. Math. Verein. 46, (1936), 179-195.

A Note on the Strong Law of Large Numbers for Partial Sums of Independent Random Vectors*

Erich Berger

Institut für Mathematische Stochastik, Universität Göttingen,

Lotzestr. 13, D-3400 Göttingen, Germany

Abstract. In 1972 Nagaev published a paper showing that necessary and sufficient conditions for the strong law of large numbers (SLLN) for independent real-valued random variables (r.v.'s) can be expressed in terms of a certain exponential estimate, applied to suitably truncated r.v.'s. This exponential estimate, in conjunction with a new majorization technique, was recently employed by the author [3] for proving a very general SLLN in the vector space setting. It is the aim of this note to demonstrate how this last mentioned result can in a convenient way be exploited for deriving more tractable (but still rather general) sufficient criteria for the SLLN.

1. Introduction and Preparatory Results

The r.v.'s occurring in this paper will always be assumed to be defined on a common probability space $(\Omega, \mathcal{A}, P)$. Convergence in probability and convergence almost surely (a.s.) are indicated by $\xrightarrow{P}$ and $\xrightarrow{a.s.}$, respectively. $I(A)$ is the indicator function of a set A; $\mathbb{R}^+ := \{x \in \mathbb{R} : x > 0\}$.

In order to describe the framework of this paper, we begin by considering a sequence $(\xi_j)_{j \in \mathbb{N}}$ of independent, symmetric, real-valued r.v.'s. Let

---

*Most of this paper has been written while the author was working at the Fachbereich Mathematik WE 1, Freie Universität Berlin.

ISBN 0-12-085520-8

$$(1.1)\qquad (a_j)_{j\in\mathbb{N}} \text{ be a nondecreasing sequence in } \mathbb{R}^+ \text{ such that } \sup\{a_j : j \in \mathbb{N}\} = \infty,$$

and write

$$(1.2)\qquad \zeta_n := \sum_{j=1}^{n} \xi_j \qquad \text{for} \quad n \in \mathbb{N}.$$

It is well-known that the question whether

$$(1.3)\qquad a_n^{-1}\zeta_n \xrightarrow{a.s.} 0$$

or not is intimately connected with the convergence properties of the series

$$(1.4)\qquad \sum_{\ell=2}^{\infty} P\{\zeta_{\nu(\ell)} - \zeta_{\nu(\ell-1)} \geq \varepsilon a_{\nu(\ell)}\}$$

for $\varepsilon \in \mathbb{R}^+$, where $(\nu(\ell))_{\ell\in\mathbb{N}}$ is a strictly increasing sequence in $\mathbb{N}$ such that $(a_{\nu(\ell)})_{\ell\in\mathbb{N}}$ grows approximately exponentially fast. Actually, it can be shown that (1.3) is equivalent to the fact that the series in (1.4) converge for each $\varepsilon \in \mathbb{R}^+$ provided $(\nu(\ell))_{\ell\in\mathbb{N}}$ is a strictly increasing sequence in $\mathbb{N}$ having the following property:

there exist positive real numbers c, d, ρ with $1 < c \leq d$ and $\rho \leq 1$ such that

$$(1.5)\qquad a_{\nu(k+r)} \geq \rho c^r a_{\nu(k)} \qquad \text{for all} \quad k,\ r \in \mathbb{N}$$

and

$$a_{\nu(\ell)} \leq d a_{\nu(\ell-1)+1} \qquad \text{for all} \quad \ell \in \mathbb{N}\setminus\{1\} \quad \text{with} \quad \nu(\ell) - \nu(\ell-1) \geq 2.$$

Given $c > 1$ and $(a_j)_{j\in\mathbb{N}}$, it is easily seen that it is always possible to find a sequence $(\nu(\ell))_{\ell\in\mathbb{N}}$ satisfying the requirements of (1.5) with $c = d = \rho^{-1}$. (For further details, see, e.g., [3], Theorem 4.1 and Lemma 4.1).

To examine the convergence properties of the series in (1.4), it is natural to introduce the truncated r.v.'s

$$(1.6)\qquad \eta_{j,\delta} := \xi_j I(\{|\xi_j| \leq \delta a_j\}) \qquad (j \in \mathbb{N},\ \delta \in \mathbb{R}^+),$$

to show that the effect of the truncation is negligible, and to estimate the probabilities

$$P\{\sum_{j=\nu(\ell)+1}^{\nu(\ell+1)} \eta_{j,\delta} \geq \varepsilon a_{\nu(\ell+1)}\} \qquad (\varepsilon \in \mathbb{R}^+,\ \ell \in \mathbb{N})$$

with the help of a suitable exponential inequality. Theorem 1.A below (a somewhat extended version of Nagaev's [15] theorem about necessary and sufficient conditions in the SLLN) provides a general justification of this approach.

To state Theorem 1.A, we need the cumulant generating functions

$$(1.7) \qquad C_{j,\delta}(t) := \log E(\exp(t\eta_{j,\delta})) \qquad (t \in \mathbb{R},\ \delta \in \mathbb{R}^+)$$

of the r.v.'s $\eta_{j,\delta}$. Furthermore, we write

$$(1.8) \qquad J(\ell) := (\nu(\ell-1),\ \nu(\ell)] \cap \mathbb{N} \qquad \text{for} \quad \ell \in \mathbb{N} \setminus \{1\}$$

and

$$(1.9) \qquad \Lambda := \{\ell \in \mathbb{N} \setminus \{1\} :\ \nu(\ell) - \nu(\ell-1) \geq 2\}.$$

Necessary and sufficient conditions for (1.3) and

$$(1.10) \qquad \limsup_{n\to\infty} a_n^{-1}\zeta_n < \infty \qquad \text{a.s.,}$$

respectively, can then be formulated by means of the condition described in the following

Definition 1.1. Notation is as above. Let $\delta,\ \varepsilon \in \mathbb{R}^+$ be fixed. We say that the sequence $(\xi_j)_{j\in\mathbb{N}}$ satisfies the ***Nagaev condition*** $N(\delta,\ \varepsilon,\ (\nu(j))_{j\in\mathbb{N}},\ (a_j)_{j\in\mathbb{N}})$ if and only if the following relations are fulfilled:

(1.11) $\sum_{j=1}^{\infty} P\{|\xi_j| > \delta a_j\} < \infty;$

(1.12) there is a sequence $(t_k)_{k\in\mathbb{N}}$ in $\mathbb{R}^+$ such that

$$\sum_{j\in J(\ell)} C'_{j,\delta}(t_\ell) \leq \varepsilon a_{\nu(\ell)} \qquad \text{for all} \quad \ell \in \Lambda$$

and

$$\sum_{\ell\in\Lambda} \exp(-\varepsilon a_{\nu(\ell)} t_\ell) < \infty.$$

The already announced theorem now reads as follows.

Theorem 1.A (essentially due to Nagaev [15]; for the present version, see [3], Theorem 4.2). *Notation is as above. Suppose the sequence* $(\nu(j))_{j\in\mathbb{N}}$ *satisfies (1.5). Then:*

(i) *(1.9) holds if and only if* $(\xi_j)_{j\in\mathbb{N}}$ *satisfies the Nagaev condition* $N(\varepsilon,\ \varepsilon,\ (\nu(j))_{j\in\mathbb{N}},\ (a_j)_{j\in\mathbb{N}})$ *for each* $\varepsilon \in \mathbb{R}^+$;

(ii) *(1.10) holds if and only if there exist* $\delta,\ \varepsilon \in \mathbb{R}^+$ *such that* $(\xi_j)_{j\in\mathbb{N}}$ *satisfies the Nagaev condition* $N(\delta,\ \varepsilon,\ (\nu(j))_{j\in\mathbb{N}},\ (a_j)_{j\in\mathbb{N}})$.

In a recent paper by the author [3], the exponential inequality underlying the proof of Theorem 1.A was combined with a new majorization technique to establish a very general almost sure convergence theorem for vector valued r.v.'s. To state this result, we need the following

Definition 1.2. Let $\mathbb{F}$ be a real vector space, let $\mathcal{F}$ be a σ-field of subsets of $\mathbb{F}$, and let $||\cdot||$ be a semi-norm on $\mathbb{F}$. The triple $(\mathbb{F},\mathcal{F},||\cdot||)$ is called a *semi-normed measurable vector space* if the following conditions are satisfied:

(1.13) addition in $\mathbb{F}$ and scalar multiplication are measurable operations ($\mathbb{F}\times\mathbb{F}\to\mathbb{F}$ and $\mathbb{R}\times\mathbb{F}\to\mathbb{F}$, respectively);

(1.14) the function $x\to ||x||$ $(x\in\mathbb{F})$ is measurable.

Now let $(\mathbb{F},\mathcal{F},||\cdot||)$ be a semi-normed measurable vector space, let $(X_j)_{j\in\mathbb{N}}$ be a sequence of independent $\mathbb{F}$-valued r.v.'s, and let $(R_j)_{j\in\mathbb{N}}$ be a sequence of independent Bernoulli r.v.'s which is also independent of $(X_j)_{j\in\mathbb{N}}$. (A r.v. $Y:\Omega\to\mathbb{R}$ is called Bernoulli if $P\{Y=1\}=P\{Y=-1\}=\frac{1}{2}$.) For $n\in\mathbb{N}$, let

$$(1.15)\qquad S_n := \sum_{j=1}^{n} X_j.$$

Finally, let

(1.16) $(a_j)_{j\in \mathbb{N}}$, $(\nu(\ell))_{\ell\in \mathbb{N}}$, c, d, ρ, $(J(\ell))_{\ell\in \mathbb{N}\setminus\{1\}}$, and Λ be as in (1.1), (1.5), (1.8), and (1.9).

Theorem 1.B (cf. [3], Theorem 4.3). *Notation is as above. Moreover, let* $\delta,\ \varepsilon,\ \gamma \in \mathbb{R}^+$, *and assume that*

(1.17) *the sequence* $(R_j||X_j||)_{j\in \mathbb{N}}$ *satisfies the Nagaev condition* $N(\delta,\ \varepsilon,\ (\nu(j))_{j\in \mathbb{N}},\ (a_j)_{j\in \mathbb{N}})$;

(1.18) $$\limsup_{n\to\infty} a_n^{-1} E||\sum_{j=1}^{n} X_j I(\{||X_j|| \le \delta a_j\})|| \le \gamma.$$

Then

(1.19) $$\limsup_{n\to\infty} a_n^{-1}||S_n|| \le d\rho^{-1}c(c-1)^{-1}\max\{8\varepsilon + 6\gamma,\ \delta\} \qquad a.s.$$

Remark 1.1. As to the question of finding sufficient conditions ensuring (1.18), we note that under the assumption

$$\sum_{j=1}^{\infty} P\{||X_j|| > \delta a_j\} < \infty,$$

implicit in (1.17), the following assertions obtain:

(i) If $a_n^{-1}||S_n|| \xrightarrow{P} 0$, then (1.18) holds with $\gamma = 8\delta$.

(ii) If the sequence $(a_n^{-1}||S_n||)_{n\in \mathbb{N}}$ is bounded in probability, then there exists a $\gamma \in \mathbb{R}^+$ such that (1.18) holds.

(For details, see [3], Lemmas 4.5 and 4.6.)

Remark 1.2. An interesting result concerning the necessity of a Nagaev type condition for the SLLN in the vector space setting has been obtained by Alt [2], Théorème 6.

To give an impression of the generality of Theorem 1.B, we mention that the following result can easily be established by combining Theorem 1.B, Theorem 1.A, and Remark 1.1.

Theorem 1.C (see [3], Theorems 1.4 and 1.5). *Notation is as above.*

(i) *If* $a_n^{-1}\sum_{j=1}^{n} R_j||X_j|| \xrightarrow{a.s.} 0$, *then* $a_n^{-1}||S_n|| \xrightarrow{a.s.} 0$ *if and only if* $a_n^{-1}||S_n|| \xrightarrow{P} 0$.

(ii) *If* $\limsup_{n\to\infty} a_n^{-1}\sum_{j=1}^{n} R_j||X_j|| < \infty$ *a.s., then* $\limsup_{n\to\infty} a_n^{-1}||S_n|| < \infty$ *a.s. if and only if the sequence* $(a_n^{-1}||S_n||)_{n\in\mathbb{N}}$ *is bounded in probability.*

It is the aim of this paper to demonstrate how Theorem 1.B can in a convenient way be exploited for deriving rather general sufficient criteria for the SLLN that are easier to deal with than (1.17). The proofs of the results given in the sequel do not depend on the hard to prove "only if" part of Theorem 1.A.

The theorems and corollaries derived in Section 2 are special cases of Theorem 1.B. The conditions used depend only on strong moments of suitably truncated r.v.'s. Section 3, which is similar in spirit to Heinkel's [7] paper, shows that our results also complement in a natural way the important recent work by Ledoux, Talagrand, and Alt ([11], [12], [1]) dealing with criteria for the SLLN that involve weak moments.

2. Stability Results for Vector Valued Random Variables

We continue to use the notation introduced in connection with Theorem 1.B above. Furthermore, let $r \in \mathbb{R}^+$, let

(2.1) $(\delta_j)_{j\in\mathbb{N}}$ be a bounded sequence in $\mathbb{R}^+$ such that

$$\sup\{\delta_j a_j : j \in J(\ell)\} = \delta_{\nu(\ell)} a_{\nu(\ell)} \text{ for all } \ell \in \Lambda,$$

define

(2.2) $\eta_j := R_j||X_j||I(\{||X_j|| \le \delta_j a_j\}) \quad \text{for} \quad j \in \mathbb{N},$

and let

(2.3) $(B_k)_{k\in \mathbb{N}}$ and $(A_k(r))_{k\in \mathbb{N}}$ be two sequences in $\mathbb{R}^+$ such that

$$B_\ell \geq \sum_{j\in J(\ell)} E(\eta_j^2) \text{ and } A_\ell(r) \geq \sum_{j\in J(\ell)} E|\eta_j|^r \text{ for all } \ell \in \mathbb{N}\setminus\{1\}.$$

To ease the intuitive understanding of the conditions appearing below, we recall that in the "classical" case $(a_j)_{j\in \mathbb{N}} = (j)_{j\in \mathbb{N}}$ one can choose $\nu(\ell) := 2^\ell$, $J(\ell) := (2^{\ell-1}, 2^\ell] \cap \mathbb{N}$, and $\Lambda = \mathbb{N}\setminus\{1\}$.

The main result of this paper is the following extension of Theorem 6 in Fuk and Nagaev [4]. (Cf. Remark 2.1 below.)

Theorem 2.1. *Let* $\varepsilon, \gamma \in \mathbb{R}^+$ *be fixed, and suppose:*

$$\sum_{j=1}^{\infty} P\{||X_j|| > \delta_j a_j\} < \infty; \tag{2.4}$$

$$\sum_{\ell\in\Lambda}\left(1+\frac{\varepsilon\delta_{\nu(\ell)}^{r-1}a_{\nu(\ell)}^r}{2A_\ell(r)}\right)^{-\varepsilon/(2\delta_{\nu(\ell)})} < \infty; \tag{2.5}$$

$$\sum_{\ell\in\Lambda}\exp(-\varepsilon^2 e^{-r} a_{\nu(\ell)}^2/(2B_\ell)) < \infty; \tag{2.6}$$

$$\limsup_{n\to\infty} a_n^{-1} E||\sum_{j=1}^{n} X_j I(\{||X_j|| \leq \delta_j a_j\})|| \leq \gamma. \tag{2.7}$$

Then

$$\limsup_{n\to\infty} a_n^{-1}||S_n|| \leq d\rho^{-1}c(c-1)^{-1}\max\{8\varepsilon + 6\gamma,\ \delta_0\} \qquad a.s., \tag{2.8}$$

where $\delta_0 := \sup\{\delta_j : j \in \mathbb{N}\}$.

The *proof* is deferred to Section 3.

Conditions (2.4)–(2.7) admit the following heuristic interpretation. (2.4) ensures that seldomly occurring very large values of the r.v.'s $||X_j||$ can be neglected. In the case of real-valued, symmetric r.v.'s X_j, the terms $\exp(-\varepsilon^2 e^{-r} a^2_{\nu(\ell)}/(2B_\ell))$ in (2.6) are closely related to the normal approximation of the probabilities

$$P\{\sum_{j\in J(\ell)} X_j I(\{|X_j| \le \delta_j a_j\}) \ge \varepsilon e^{-r/2} a_{\nu(\ell)}\}.$$

(Note that

$$(2.9) \qquad (2^{1/2}-1)e^{-x^2} \le \int_x^\infty e^{-t^2/2}dt \le e^{-x^2/2}$$

for all $x \ge 1$.) The terms of the series in (2.5) serve the purpose of controlling the error term of this approximation. If nontrivial information about r-th order absolute moments is available, it is usually advantageous to apply Theorem 2.1 with a large value of r. Condition (2.7) is connected with the stochastic boundedness (or stochastic convergence) properties of the sequence $(a_n^{-1}||S_n||)_{n\in\mathbb{N}}$ (see Remark 1.1 above). Under additional assumptions about the geometric structure of the space $(\mathbb{F}, ||\cdot||)$, this condition can be replaced by a centering condition (see Theorem 2.2 below).

Before stating Theorem 2.2, we recall that a real separable Banach space $(\mathbb{B}, ||\cdot||_{\mathbb{B}})$ is said to be of *type 2* if and only if there is a constant $q_{\mathbb{B}} \in \mathbb{R}^+$ such that for any finite sequence $Y_1, \ldots, Y_n$ of independent $\mathbb{B}$-valued r.v.'s with

$$(2.10) \qquad E||Y_j||^2_{\mathbb{B}} < \infty \quad \text{and} \quad EY_j = 0 \quad \text{for all} \quad j \in \{1,\ldots,n\}$$

one has

$$(2.11) \qquad E||\sum_{j=1}^n Y_j||^2_{\mathbb{B}} \le q_{\mathbb{B}} \sum_{j=1}^n E||Y_j||^2_{\mathbb{B}}.$$

(For further details, see, e.g., [8].) If $(\mathbb{F}, ||\cdot||)$ is a type 2 Banach space (e.g., an L_p-space with $2 \le p < \infty$), the following version of Theorem 2.1 holds.

Theorem 2.2. *Notation is as in Theorem 2.1. In case $(\mathbb{F}, ||\cdot||)$ is a real separable Banach space of type 2 and $\mathcal{F}$ is its Borel σ-field, the assertion of Theorem 2.1 remains true if (2.7) is replaced by*

$$(2.12) \qquad E(X_j I(\{||X_j|| \le \delta_j a_j\})) = 0 \qquad \textit{for all} \quad j \in \mathbb{N},$$

provided γ *is chosen to be equal to* $q_F^{1/2} d\rho^{-1} c(c^2-1)^{-1/2}\delta_0$ *(cf. (1.16) and (2.11)).*

Proof. For $j \in \mathbb{N}$, let

$$Z_j := X_j I(\{||X_j|| \le \delta_j a_j\}).$$

Also define

$$U_n := \sum_{j=1}^{n} Z_j \qquad \text{for} \quad n \in \mathbb{N}.$$

It suffices to prove that

$$\limsup_{n\to\infty} a_n^{-1} E||U_n|| \le q_F^{1/2} d\rho^{-1} c(c^2-1)^{-1/2}\delta_0. \tag{2.13}$$

Fix $\ell \in \mathbb{N}\setminus\{1\}$ and $m \in J(\ell)$. Setting $\nu(0) := 0$, we have

$$E||U_m||^2 \le q_F \sum_{k=1}^{\ell} \sum_{j=\nu(k-1)+1}^{\nu(k)} E||Z_j||^2 \qquad \text{(by (2.12))}. \tag{2.14}$$

(2.6) implies that the sequence $(a_{\nu(\ell)}^{-2} B_\ell)_{\ell\in\Lambda}$ converges to 0 as $\ell \to \infty$. Hence and from the inequality $||Z_j|| \le \delta_j a_j$ (for all $j \in \mathbb{N}$), we infer that

$$E||U_m||^2 \le q_F \sum_{k=1}^{\ell} \theta_k a_{\nu(k)}^2, \tag{2.15}$$

where $(\theta_k)_{k\in\mathbb{N}}$ is a sequence in $\mathbb{R}^+$ with

$$\limsup_{k\to\infty} \theta_k \le \delta_0^2. \tag{2.16}$$

Taking (1.5) into account, it follows from (2.15) that

$$\begin{aligned} a_m^{-2} E||U_m||^2 &\le q_F d^2 \sum_{k=1}^{\ell} \theta_k (a_{\nu(k)}^2 a_{\nu(\ell)}^{-2}) \\ &\le q_F d^2 \rho^{-2} \sum_{k=1}^{\ell} c^{-2(\ell-k)} \theta_k. \end{aligned}$$

Together with (2.16) and the Toeplitz lemma, this entails that

$$\limsup_{n\to\infty} a_n^{-2} E||U_n||^2 \le q_F d^2 \rho^{-2} c^2 (c^2-1)^{-1} \delta_0^2,$$

which implies (2.13).

Remark 2.1. In the case $\mathbb{F} = \mathbb{R}$, $a_j = j$ for all $j \in \mathbb{N}$, Theorem 2.2 is essentially a variant of Theorem 6 in Fuk and Nagaev [4].

To illustrate the scope of Theorems 2.1 and 2.2, we shall now state and prove several corollaries to these theorems.

Corollary 2.1. *The assertions of Theorems 2.1 and 2.2 remain true if (2.5) and (2.6) are replaced by*

$$\sum_{\ell\in\Lambda}\left(1+\frac{\varepsilon e^{-2}\delta_{\nu(\ell)}a^2_{\nu(\ell)}}{B_\ell}\right)^{-\varepsilon/(2\delta_{\nu(\ell)})}<\infty. \tag{2.17}$$

Proof. Choose $r = 2$ and $B_\ell = A_\ell(r)$ in (2.5) and (2.6). Then it is obvious that (2.17) implies (2.5); moreover, the inequality $\log(1 + x) \leq x$ (valid for all $x \in \mathbb{R}^+$) shows that (2.17) also implies (2.6).

Remark 2.2 (conditions that are "essentially equivalent" to (2.17)). Setting

$$\tau(\ell) := \varepsilon/\delta_{\nu(\ell)} \quad \text{and} \quad \beta(\ell) := \varepsilon\delta_{\nu(\ell)}a^2_{\nu(\ell)}/B_\ell \quad \text{for} \quad \ell \in \Lambda, \tag{2.18}$$

(2.17) takes the form

$$\sum_{\ell\in\Lambda}(1+e^{-2}\beta(\ell))^{-\tau(\ell)/2}<\infty. \tag{2.19}$$

Prohorov [18] and Wittmann [20] have utilized variants of the conditions

$$\sum_{\ell\in\Lambda}\exp(-\tau(\ell)\text{ arc sinh }\beta(\ell))<\infty \tag{2.20}$$

and

$$\sum_{\ell\in\Lambda}\exp(-\frac{\kappa(\ell)}{2})<\infty, \tag{2.21}$$

where $\kappa(\ell)$ is the solution of the equation $(\kappa(\ell)/\tau(\ell))\exp(\kappa(\ell)/\tau(\ell)) = \beta(\ell)$, for proving stability results for real-valued r.v.'s. To compare (2.19)–(2.21), we first notice that, for all $y \in \mathbb{R}^+$,

$$\text{arc sinh } y = \log(y + (1 + y^2)^{1/2}), \tag{2.22}$$

$$(2.23)\qquad (1+y)\log(1+y) \geq y,$$

and (see, e.g., Mitrinović [14], p. 272)

$$(2.24)\qquad \frac{1}{2}(1+y)^{1/2}\log(1+y) \leq y.$$

Using (2.22), we see that

$$(2.25)\qquad (1+2\beta(\ell))^{-\tau(\ell)} \leq \exp(-\tau(\ell)\ \text{arc sinh}\ \beta(\ell)) \leq (1+\beta(\ell))^{-\tau(\ell)}$$

for all $\ell \in I\!N$. Applying (2.23) and (2.24), we get

$$(2.26)\qquad (1+\beta(\ell))^{-\tau(\ell)/2} \leq \exp(-\frac{\kappa(\ell)}{2}) \leq (1+\beta(\ell))^{-\tau(\ell)/4} \qquad \text{for all} \quad \ell \in I\!N.$$

The inequalities (2.25) and (2.26) lead to the effect that the conditions (2.19), (2.20), and (2.21) are in a sense "almost equivalent". Indeed, multiplying the a_k's by a suitable factor and changing ε, we find that the different form of the conditions (2.19)–(2.21) can only alter the value of the constant on the right-hand side of (2.8), but not the qualitative statement. In conjunction with Remark 1.1, Corollary 2.1 in particular implies Heinkel's [7] Theorem 2.1, which in turn is an extension to vector valued r.v.'s of the sufficiency part of Prohorov's [18] Theorem 2. It can also be regarded as a vector space analogue of Wittmann's [20] Theorem 3.3.

The next corollary should be compared with the Alt-Ledoux-Talagrand theorem (Theorem 3.A below).

Corollary 2.2. *The assertions of Theorems 2.1 and 2.2 remain true if (2.5) and (2.6) are replaced by the assumption that there exists a constant* $q \in I\!R^+$ *such that*

$$(2.27)\qquad \sum_{\ell \in \Lambda} \exp(-\varepsilon q/(2\delta_{\nu(\ell)})) < \infty$$

and

$$(2.28)\qquad \sum_{\ell \in \Lambda} \exp(-\varepsilon^2 q e^{-2} a^2_{\nu(\ell)}/(2(e^q-1)B_\ell)) < \infty.$$

Proof. In view of Corollary 2.1, we need only show that (2.17) is implied by (2.27) and (2.28). But this is immediate from the inequality $\log(1+x) \geq q\min\{1, x/(e^q-1)\}$ valid for all $x \in \mathbb{R}^+$.

Remark 2.3. In the case of real-valued r.v.'s, Corollary 2.2 contains the sufficiency part of Prohorov's [18] Theorem 1 and its extension to arbitrary norming sequences given by Wittmann [20], Corollary 3.9. Together with the results quoted in Remark 1.1, it also covers the vector space analogue of Prohorov's SLLN established by Kuelbs and Zinn [10], Theorem 2, and, except for the value of the constant in the conclusion, Kuelbs' [9] vector space analogue of the upper class part of Kolmogorov's law of the iterated logarithm.

Corollary 2.3. *In case $\delta_j = \delta_0$ for all $j \in \mathbb{N}$, the assertions of Theorems 2.1 and 2.2 remain true if (2.5) is replaced by*

$$\sum_{\ell\in\Lambda}(a_{\nu(\ell)}^{-r}A_\ell(r))^{\varepsilon/(2\delta_0)} < \infty. \tag{2.29}$$

Proof. Obvious.

Corollary 2.4. *In case $\varepsilon = 2\delta_0$ and $\delta_j = \delta_0$ for all $j \in \mathbb{N}$, the assertions of Theorems 2.1 and 2.2 remain true if (2.5) is replaced by*

$$\sum_{k=1}^{\infty} a_k^{-r}E|\eta_k|^r < \infty. \tag{2.30}$$

Proof. Obvious from (1.5) and Corollary 2.3.

Remark 2.4. Taking (2.9) into account and using the symmetrization argument described in Stout's [19] book (p. 116 ff.), we see that the case $r = 3$ of the above corollary covers the sufficiency part of Martikainen's [13] Theorem 6, where real-valued r.v.'s are considered.

Corollary 2.5. *Notation and basic assumptions are as in Theorems 2.1 and 2.2, respectively. Suppose the X_j's form a sequence of independent, identically distributed* *r.v.'s, the a_j's satisfy an additional growth condition of the form*

$$(2.31) \qquad \sum_{k=n}^{\infty} a_k^{-r} = O(na_n^{-r}),$$

and that $\varepsilon = 2\delta_0$ and $\delta_j = \delta_0$ for all $j \in \mathbb{N}$. Then the assertions of Theorems 2.1 and 2.2 remain true if condition (2.5) is omitted.

Proof. In view of Corollary 2.4, it is enough to show that under the above assumptions (2.30) is automatically fulfilled. Following the reasoning in Petrov's [17] book (p. 275), we find that (with $a_0 := 0$ and for a suitable constant $c \in \mathbb{R}^+$)

$$\begin{aligned}\sum_{k=1}^{\infty} a_k^{-r} E|\eta_k|^r &= \sum_{k=1}^{\infty} a_k^{-r} \int_{\{||X_1|| \leq \delta_0 a_k\}} ||X_1||^r \, dP \\ &\leq c\delta_0^r \sum_{k=0}^{\infty} P\{||X_1|| > \delta_0 a_k\} \\ &< \infty \qquad\qquad \text{(by (2.4))}.\end{aligned}$$

Remark 2.5. (a) It is easy to check that (2.31) is satisfied if there is a $\theta \in \mathbb{R}^+$ such that the sequence $(a_n/n^{\theta+1/r})_{n\in\mathbb{N}}$ is monotonically increasing.

(b) For $a_n^{-1}||S_n|| \xrightarrow{a.s.} 0$ it is necessary that the condition

$$(2.32) \qquad \sum_{k=1}^{\infty} P\{||X_1|| > \delta_0 a_k\} < \infty$$

be satisfied for any $\delta_0 \in \mathbb{R}^+$. Hence and from (2.9) and the symmetrization argument already mentioned in Remark 2.4, it follows that the case $r = 3$ of Corollary 2.5 can be regarded as a vector space analogue of a variant of the sufficiency part of Martikainen's [13] Theorem 1.

Corollary 2.6. *In the case $\delta_j = \delta_0$ for all $j \in \mathbb{N}$, the assertions of Theorems 2.1 and 2.2 remain true if (2.5) and (2.6) are replaced by*

$$(2.33) \qquad \sum_{\ell\in\Lambda} (a_{\nu(\ell)}^{-2} B_\ell)^{\varepsilon/(2\delta_0)} < \infty.$$

Proof. Obvious from Corollary 2.1.

Remark 2.6. In the case of real-valued r.v.'s, Corollary 2.6 easily implies Corollary 2 in Section 6 of [4] and Corollary 3.4 in [20]. In conjunction with Remark 1.1, Corollary 2.6 also covers Theorem 1 of Kuelbs and Zinn [10] (where there is no essential loss of generality in assuming that $p = 2$).

3. Possible Extensions

Now let $(\mathbb{F}, ||\cdot||)$ be a real separable Banach space with topological dual $\mathbb{F}^*$. The assumptions underlying the theorems and corollaries stated in the preceding section only relate to strong moments. In recent years it has turned out that it is often more appropriate to work with conditions about weak moments (see, e.g., [1], [5], [6], [7], [11], [12]). One of the most striking results in this respect is the following extension of Prohorov's [18] Theorem 1 to $\mathbb{F}$-valued r.v.'s.

Theorem 3.A (cf. Alt [1], Théorème 7; Ledoux and Talagrand [11], Theorem 13). *Let* $(Y_j)_{j\in\mathbb{N}}$ *be a sequence of independent* $\mathbb{F}$*-valued r.v.'s, and let* $(a_j)_{j\in\mathbb{N}}$, $(\nu(\ell))_{\ell\in\mathbb{N}}$, c, d, ρ, $(J(\ell))_{\ell\in\mathbb{N}\setminus\{1\}}$, *and* Λ *be as in (1.1), (1.5), (1.8), and (1.9). Moreover, let* $(\delta_j^*)_{j\in\mathbb{N}}$ *be a null sequence in* $\mathbb{R}^+$ *such that*

$$(3.1) \qquad \sup\{\delta_j^* a_j : j \in J(\ell)\} = \delta_{\nu(\ell)}^* a_{\nu(\ell)}.$$

Suppose

$$(3.2) \qquad ||Y_j|| \le \delta_j^* a_j \qquad a.s. \text{ for all } j \in \mathbb{N};$$

$$(3.3) \qquad EY_j = 0 \qquad \text{for all } j \in \mathbb{N};$$

$$(3.4) \qquad \text{there is a constant } q \in \mathbb{R}^+ \text{ such that } \sum_{\ell\in\Lambda} \exp(-q/\delta_{\nu(\ell)}^*) < \infty.$$

For $\ell \in \mathbb{N} \setminus \{1\}$, *define*

$$(3.5) \qquad B_\ell^* := \sup\{\sum_{j\in J(\ell)} Ef^2(Y_j) : f \in \mathbb{F}^*, ||f|| \le 1\}.$$

Then, writing $T_n := \sum_{j=1}^{n} Y_j$ *for* $n \in \mathbb{N}$, *one has*

$$(3.6) \qquad a_n^{-1} T_n \xrightarrow{a.s.} 0 \qquad \textit{as} \quad n \to \infty$$

if and only if

$$(3.7) \qquad a_n^{-1} T_n \xrightarrow{P} 0 \qquad \textit{as} \quad n \to \infty$$

and

$$(3.8) \qquad \sum_{\ell \in \Lambda} \exp(-\theta a_{\nu(\ell)}^2 / B_\ell^*) < \infty \qquad \textit{for any} \quad \theta \in \mathbb{R}^+.$$

Remark 3.1. Actually, the formulation of Théorème 7 in [1] is slightly different. However, taking the remarks in connection with (1.4) and (1.5) (above) into account, it is easy to check that the proof in [1] remains applicable to establish the present result.

There is a useful double truncation type argument that makes it possible to combine Theorem 3.A and results of the form described in Section 2 above in order to relax the somewhat restrictive boundedness conditions (3.2) and (3.4) in the sufficiency part of Theorem 3.A. (See also Heinkel [7].) To describe this method, let $(Z_j)_{j \in \mathbb{N}}$ be a sequence of independent $\mathbb{F}$-valued r.v.'s, and let $(a_j)_{j \in \mathbb{N}}$, $(\delta_j)_{j \in \mathbb{N}}$, and $(\delta_j^*)_{j \in \mathbb{N}}$ be as in (1.1), (2.1), and (3.1), respectively. Also assume that $\delta_j^* \leq \delta_j$ for all $j \in \mathbb{N}$. For simplicity, suppose

$$(3.9) \qquad ||Z_j|| \leq \delta_j a_j \quad \text{a.s. and} \quad EZ_j = 0 \quad \text{for all} \quad j \in \mathbb{N}.$$

Then the r.v.'s Z_j can be decomposed in the form

$$Z_j = Y_j + X_j,$$

where

$$Y_j := Z_j' - EZ_j' \qquad \text{and} \qquad X_j := Z_j'' - EZ_j''$$

with

$$Z'_j = Z_j I(\{\|Z_j\| \leq \delta^*_j a_j\}) \qquad \text{and} \qquad Z''_j := Z_j - Z'_j.$$

Next we consider the corresponding partial sums

$$U_n := \sum_{j=1}^{n} Z_j, \quad T_n := \sum_{j=1}^{n} Y_j, \quad S_n := \sum_{j=1}^{n} X_j \qquad (n \in \mathbb{N}).$$

In order to prove that

$$a_n^{-1} U_n \xrightarrow{a.s.} 0,$$

it suffices to show that

$$a_n^{-1} T_n \xrightarrow{a.s.} 0 \qquad \text{and} \qquad a_n^{-1} S_n \xrightarrow{a.s.} 0.$$

The convergence of $(a_n^{-1}T_n)_{n\in\mathbb{N}}$ can be investigated by means of Theorem 3.A, and the convergence of the sequence $(a_n^{-1}S_n)_{n\in\mathbb{N}}$ can be examined by using the results of the present paper (Theorem 1.B, Theorem 2.1, Theorem 2.2, and Corollaries 2.1–2.6). This approach actually seems to lead to rather sharp conditions. For instance, Corollary 2.1 above does not remain true if the definition of the B_ℓ's is modified by analogy with (3.5) (cf. Heinkel [7], p. 120 ff.; recall Remark 2.2). The results that can be obtained by applying the just indicated device in particular cover Heinkel's [7] Theorem 2.2.

4. Proof of Theorem 2.1

We preface the proof of Theorem 2.1 by two lemmas.

Lemma 4.1. *Let* $\alpha \in \mathbb{R}^+$, *and let* Y *be a real-valued r.v. such that*

(4.1) $\qquad E(e^{tY}) < \infty \qquad$ *for all* $t \in (-\alpha, \alpha)$.

For $t \in (-\alpha, \alpha)$, *put* $R(t) := E(e^{tY})$ *and* $C(t) := \log R(t)$. *If* $EY = 0$, *then*

(4.2) $\qquad C'(t) \leq R'(t) \qquad$ *for all* $t \in (-\alpha, \alpha)$.

Proof. We have

$$R(0) = 1, \;\; R'(0) = EY = 0, \;\; \text{and, for all}$$
$$t \in (-\alpha, \alpha), \;\; R''(t) = E(Y^2 e^{tY}) \geq 0,$$

which implies that

$$C'(t) = R'(t)/R(t) \leq R'(t) \qquad \text{for all} \quad t \in (-\alpha, \alpha).$$

Lemma 4.2. *Let* $b, \, r \in \mathbb{R}^+$, *and let* Y *be a real-valued r.v. such that*

$$|Y| \leq b \quad \textit{a.s. and} \quad EY = 0. \tag{4.3}$$

Put $\sigma^2 := EY^2$ *and* $\tau_r := E|Y|^r$. *Then*

$$E(Ye^{tY}) \leq te^r \sigma^2 + tb^{2-r} e^{tb} \tau_r \qquad \textit{for all} \quad t \in \mathbb{R}^+. \tag{4.4}$$

Proof. The proof is similar to that of Lemma 1.4 in Nagaev's [16] article. For any $t \in \mathbb{R}^+$, we have

$$\begin{aligned}
E(Ye^{tY}) &= E(Y(e^{tY} - 1)) \\
&\leq tE(Y^2 e^{t|Y|}) \\
&= t \int_{\{|Y| \leq r/t\}} Y^2 e^{t|Y|} \, dP + t \int_{\{|Y| > r/t\}} Y^2 e^{t|Y|} \, dP \\
&\leq te^r \sigma^2 + t^{r+1} \int_{\{|Y| > r/t\}} \frac{Y^2 e^{t|Y|}}{t^r |Y|^r} |Y|^r \, dP \\
&\leq te^r \sigma^2 + t^{r+1} \frac{b^2 e^{tb}}{t^r b^r} \tau_r \\
&= te^r \sigma^2 + tb^{2-r} e^{tb} \tau_r.
\end{aligned}$$

Proof of Theorem 2.1. Because of (2.4) and the Borel-Cantelli lemma, we may, without loss of generality, assume that

$$||X_j|| \leq \delta_j a_j \qquad \text{a.s. for all} \quad j \in \mathbb{N}. \tag{4.5}$$

Taking (2.7) into account, it follows from Theorem 1.B that we need only verify that the sequence $(t_k)_{k\in\mathbb{N}}$ defined by

$$t_k := \frac{1}{2}\min\left\{\frac{\varepsilon a_{\nu(k)}}{e^r B_k}, \frac{1}{\delta_{\nu(k)}a_{\nu(k)}}\log\left(1+\frac{\varepsilon\delta_{\nu(k)}^{r-1}a_{\nu(k)}^r}{2A_k(r)}\right)\right\} \tag{4.6}$$

satisfies the conditions in (1.12). It is obvious from (2.5) and (2.6) that the second condition in (1.12) is fulfilled. To validate the first one, we define

$$R_j(t) := E(\exp(t\eta_j)) \qquad \text{and} \qquad C_j(t) := \log R_j(t)$$

for $t \in \mathbb{R}$ and $j \in \mathbb{N}$. In view of Lemma 4.1, it suffices to show that

$$\sum_{j\in J(\ell)} R_j'(t_\ell) \le \varepsilon a_{\nu(\ell)} \qquad \text{for all} \quad \ell \in \Lambda. \tag{4.7}$$

Applying Lemma 4.2, we get

$$\sum_{j\in J(\ell)} R_j'(t_\ell) = \sum_{j\in J(\ell)} E(\eta_j \exp(t_\ell\eta_j)) \le T_1 + T_2, \tag{4.8}$$

where

$$T_1 := t_\ell e^r B_\ell \qquad \text{and} \qquad T_2 := t_\ell\delta_{\nu(\ell)}^{2-r}a_{\nu(\ell)}^{2-r}\exp(\delta_{\nu(\ell)}a_{\nu(\ell)}t_\ell)A_\ell(r)$$

(cf. (2.1)). It is obvious that

$$T_1 \le \frac{1}{2}\varepsilon a_{\nu(\ell)}. \tag{4.9}$$

In order to demonstrate that

$$T_2 \le \frac{1}{2}\varepsilon a_{\nu(\ell)}, \tag{4.10}$$

let $y := \varepsilon a_{\nu(\ell)}/(2\delta_{\nu(\ell)}^{1-r}a_{\nu(\ell)}^{1-r}A_\ell(r))$. The inequality (2.24) can be written in the form

$$\frac{1}{2}\log(1+y)\exp(\frac{1}{2}\log(1+y)) \le y.$$

Recalling the definition of y and t_ℓ, this entails that

$$t_\ell\delta_{\nu(\ell)}a_{\nu(\ell)}\exp(\delta_{\nu(\ell)}a_{\nu(\ell)}t_\ell) \le \varepsilon a_{\nu(\ell)}/(2\delta_{\nu(\ell)}^{1-r}a_{\nu(\ell)}^{1-r}A_\ell(r)),$$

giving (4.10). Combining (4.8)–(4.10), we arrive at (4.7).

Acknowledgement. Thanks to the referee for some useful comments. His remarks in particular led to the incorporation of Section 3.

References

1. Alt, J.-C.: Une forme générale de la loi forte des grands nombres pour des variables aléatoires vectorielles. In: Cambanis, S., Weron, A. (eds.) Probability Theory on Vector Spaces IV. Proceedings Łańcut 1987. (Lect. Notes Math., vol. 1391, pp. 1–15) Berlin Heidelberg New York: Springer 1989

2. Alt, J.-C.: Sur la loi des grands nombres de Nagaev en dimension infinie. In: Probability in Banach Spaces VII. Proc. Conf. Oberwolfach 1988. Boston Basel Stuttgart: Birkhäuser 1990

3. Berger, E.: Majorization, exponential inequalities, and almost sure behaviour of vector valued random variables. To appear in Ann. Probab.

4. Fuk, D. Kh., Nagaev, S. V.: Probability inequalities for sums of independent random variables. Theory Probab. Appl. **16**, 643–660 (1971)

5. Heinkel, B.: On the law of large numbers in 2-uniformly smooth Banach spaces. Ann. Prob. **12**, 851–857 (1984)

6. Heinkel, B.: Une extension de la loi des grands nombres de Prohorov. Z. Wahrsch. verw. Gebiete **67**, 349–362 (1984)

7. Heinkel, B.: A law of large numbers for random vectors having large norms. In: Probability in Banach Spaces VII. Proc. Conf. Oberwolfach 1988. Boston Basel Stuttgart: Birkhäuser 1990

8. Hoffmann-Jørgensen, J., Pisier, G.: The law of large numbers and the central limit theorem in Banach spaces. Ann. Probab. **4**, 587–599 (1976)

9. Kuelbs, J.: Kolmogorov's law of the iterated logarithm for Banach space valued random variables. Illinois J. Math. **21**, 784–800 (1977)

10. Kuelbs, J., Zinn, J.: Some stability results for vector valued random variables. Ann. Probab. **7**, 75–84 (1979)

11. Ledoux, M., Talagrand, M.: Comparison theorems, random geometry and some limit theorems for empirical processes. Ann. Probab. **17**, 596–631 (1989)

12. Ledoux, M., Talagrand, M.: Some applications of isoperimetric methods to strong limit theorems for sums of independent random variables. Ann. Prob. **18**, 754–789 (1990)

13. Martikainen, A. I.: Criteria for strong convergence of normalized sums of independent random variables and their applications. Theory Probab. Appl. **29**, 519–533 (1984)

14. Mitrinović, D. S.: Analytic inequalities. Berlin Heidelberg New York: Springer 1970

15. Nagaev, S. V.: On necessary and sufficient conditions for the strong law of large numbers. Theory Probab. Appl. **17**, 573–581 (1972)

16. Nagaev, S. V.: Large deviations of sums of independent random variables. Ann. Probab. **7**, 745–789 (1979)

17. Petrov, V. V.: Sums of independent random variables. Berlin Heidelberg New York: Springer 1975

18. Prohorov, Yu. V.: Some remarks on the strong law of large numbers. Theory Probab. Appl. **4** , 204–208 (1959)

19. Stout, W.: Almost Sure Convergence. New York: Academic Press 1974

20. Wittmann, R.: Sufficient moment and truncated moment conditions for the law of the iterated logarithm. Probab. Th. Rel. Fields **75**, 509–530 (1987)

SUMMABILITY METHODS AND ALMOST-SURE CONVERGENCE

N.H. Bingham and L.C.G. Rogers

§0. Introduction

This paper explores links between probability theory and summability theory. Such links are to be expected, since a summability method is essentially a (limit of) a weighted average, while the use of weighted averages - be they expectations, sample means, or variants thereof - is ubiquitous in probability and statistics.

The paper falls into two parts. In §1, we present three results (Theorems 1-3) on limits of occupation times (and for comparison, a result of Brosamler, Theorem 4), the theme being the interplay between density properties of sets and limiting properties of occupation times of sets by random processes. In §§2-4, we survey the general area of links between probability and summability, focusing particularly on the i.i.d. case, and comparing the strengths of the integrability conditions on the distribution and the summability method in the a.s. convergence statement.

To make the paper self-contained, we review here the summability methods that appear below. For background, see e.g. Hardy (1949).

Cesàro methods $C_\alpha, \alpha > 0$: $s_n \to s\ (C_\alpha)$ means

$$\frac{1}{A_n^\alpha}\sum_{k=0}^{n} A_{n-k}^{\alpha-1} s_k \to s \quad (n \to \infty); \quad A_n^\alpha := (\alpha+1)\ldots(\alpha+n)/n!\,.$$

Abel method A: $s_n \to s\ (A)$ means

$$(1-r)\sum_{k=0}^{\infty} s_k r^k \to s \quad (r \uparrow 1).$$

Riesz method $R(\lambda_n, k)$ of order k based on $\lambda_n \uparrow \infty$: $s_n \to s \quad (R, \lambda_n, k)$ means

$$\frac{k}{x^k}\int_0^x A_\lambda(t)(x-t)^{k-1}dt \to s \quad (x \to \infty)$$

$(A_\lambda(x) := s_n$ for $\lambda_n < x \le \lambda_{n+1})$.

Euler method $E(\lambda), \lambda > 0$: $s_n \to s\ (E(\lambda))$ means

$$(1+\lambda)^{-n}\sum_{k=0}^{n} s_k \binom{n}{k}\lambda^k \to s \quad (n \to \infty).$$

Borel method B: $s_n \to s\ (B)$ means

$$e^{-\lambda}\sum_{k=0}^{\infty} s_k.\lambda^k/k! \to s \quad (\lambda \to \infty).$$

ISBN 0-12-085520-8

§1. Limits of occupation times

The theme of the results of this section is the use of summability methods to link density properties of sets with limit behaviour of occupation times of these sets by random processes.

We begin by considering sets $A \subset \mathbf{R}^+$; the process will be Brownian motion on $\mathbf{R}$ with unit drift,

$$X_t = B_t + t$$

with B standard Brownian motion; the summability method will be the Cesàro method; $|\cdot|$ denotes Lebesgue measure.

THEOREM 1. *A set $A \subset \mathbf{R}^+$ has Cesàro density c,*

$$\frac{1}{t}|A \cap [0,t]| \to c \quad (t \to \infty),$$

if and only if its occupation time by drifting Brownian motion X satisfies the strong law

$$\frac{1}{t}\int_0^t I(X_u \in A)du \to c \quad (t \to \infty) \quad \text{a.s.}$$

The random-walk analogue of this result is due to Stam (1968) and Meilijson (1973); for extensions see Bingham and Goldie (1982), Högnäs and Mukherjea (1984), Berbee (1987).

Proof. Drifting Brownian motion X is a Lévy process with Lévy exponent $\psi(s) = s + \frac{1}{2}s^2$. Its first-passage process $\tau = (\tau_u)_{u>0}$, where

$$\tau_u := \inf\{t : X_t > u\}, \qquad \tau_0 := 0$$

is a subordinator as X is spectrally negative. Its Lévy exponent $\eta(s)$ satisfies $\psi(\eta(s)) = s$ (see e.g. Bingham (1975), §4), so

$$\eta(s) = -1 + (1+2s)^{\frac{1}{2}}.$$

Thus $E\tau_1 = \eta'(0) = 1$, $E\tau_u = u$, and by the strong law

$$\tau_u/u \to 1 \quad \text{a.s.} \quad (u \to \infty).$$

Write

$$\xi_n := \int_{\tau_{n-1}}^{\tau_n} I_A(X_u)du, \quad \mu_n := E\xi_n, \quad \tilde{\xi}_n := \xi_n - \mu_n.$$

Then the $\tilde{\xi}_n$ are independent zero-mean random variables with

$$\text{var}\,\tilde{\xi}_n \le E\xi_n^2 \le E(\tau_n - \tau_{n-1})^2 = E\tau_1^2 < \infty,$$

so the $\tilde{\xi}_n$ are bounded in L_2. The martingale

$$M_n := \sum_1^n \tilde{\xi}_j / j$$

is thus bounded in L_2, so almost-surely convergent. By Kronecker's lemma, this gives

$$\frac{1}{n}\sum_1^n \tilde{\xi}_k = \frac{1}{n}\sum_1^n \xi_k - \frac{1}{n}\sum_1^n \mu_k \to 0 \quad \text{a.s.} \tag{1}$$

Write $(L(t,x))$ for the local time of X, jointly continuous in t and x by Trotter's theorem (see Rogers and Williams (1987), 101). Then

$$\begin{aligned} \sum_1^n \mu_k &= E\int_0^{\tau_n} I_A(X_u)du \\ &= E\int_0^n I_A(x)L(\tau_n,x)dx \\ &= \int_0^n I_A(x)EL(\tau_n,x)dx. \end{aligned} \tag{2}$$

To compute $EL(\tau_n,x)$, we use Tanaka's formula:

$$X_{\tau_k}^- = -\int_0^{\tau_k} I_{\{X_u \le 0\}} d(B_u + u) + \tfrac{1}{2}L(\tau_k,0).$$

(Rogers and Williams (1987), IV.43.6). When we take expectations, the stochastic integral with respect to B contributes nothing, since integrability of τ_k implies L^2-boundedness of the stochastic integral $\int_0^{t\wedge\tau_k} I_{\{X_u\le 0\}}dB_u$. Since $X_{\tau_k}^- = 0$, we deduce that

$$\begin{aligned} EL(\tau_k,0) &= 2E\int_0^{\tau_k} I(X_u \le 0)du \\ &= 2E\Big(\int_0^\infty - \int_{\tau_k}^\infty\Big)I(X_u \le 0)du. \end{aligned}$$

Now the all-time minimum of X_t is exponentially distributed with parameter 2 (see e.g. Bingham (1975), Prop. 5b applied to $-X$):

$$P(X_t \le 0 \text{ for some } t\,|X_0 = k) = e^{-2k}.$$

Using this and the strong Markov property at time τ_k,

$$\begin{aligned} EL(\tau_k,0) &= 2(1-e^{-2k})E\int_0^\infty I(X_u \le 0)du \\ &= c(1-e^{-2k}), \text{ say.} \end{aligned}$$

Similarly,

$$EL(\tau_k,x) = c(1-e^{-2k(k-x)}) \qquad (0 < x < k).$$

The constant $c = 2\int_0^\infty P(X_u \le 0)du$ is easily evaluated by simple calculus to be 1.

Hence by (2),

$$\sum_1^n \mu_k = \int_0^n I_A(x)\ (1 - e^{-2(n-x)})dx,$$

so

$$\sum_1^n \mu_k \le \int_0^n I_A(x)dx \le \sum_1^n \mu_k + \int_0^n e^{-2(n-x)}dx \le \sum_1^n \mu_k + \tfrac{1}{2}.$$

In particular,

$$\frac{1}{n}\sum_1^n \mu_k - \frac{1}{n}\int_0^n I_A(x)dx \to 0. \tag{3}$$

Combining (1) and (3),

$$\frac{1}{n}\int_0^{\tau_n} I_A(X_u)du - \frac{1}{n}\int_0^n I_A(x) \to 0 \quad \text{a.s.}$$

Now $\tau_t/t \to 1$ a.s., and the integrands are bounded. Hence

$$\frac{1}{t}\int_0^t I_A(X_u)du - \frac{1}{t}\int_0^t I_A(x)dx \to 0 \quad \text{a.s.} \quad (t \to \infty),$$

which proves the result, and more. In fact, Theorem 1 is of equi-convergence rather than convergence character: the difference above converges though neither term need do so. This is to be expected, in view of the similar nature of the random-walk result (Bingham and Goldie (1982), Theorems 1,2,2′).

Use of Trotter's theorem in a similar context may be found in Kendall and Westcott (1987), Theorem 6.7.

When Theorem 1 applies,

$$\frac{1}{t}\int_0^t P(X_u \in A)du \to c \quad (t \to \infty):$$

$$P(X_t \in A) \to c \quad \text{in the Cesàro sense.}$$

If we ask instead for pointwise convergence here, we need A to have density c in a sense correspondingly stronger than the Cesàro sense:

THEOREM 2.

(i) $$P(X_t \in A) \to c \qquad (t \to \infty)$$

if and only if

(ii) $$\frac{1}{u\sqrt{t}}|A \cap [t, t + u\sqrt{t}]| \to c \quad (t \to \infty) \quad \textit{for all } u > 0.$$

Proof. Statement (i) is

$$\frac{1}{(2\pi t)^{\frac{1}{2}}}\int_0^\infty I_A(y)\exp\{-\tfrac{1}{2}(t-y)^2/t\}dy \to c \quad (t\to\infty),$$

or

$$I_A(x)\to c \qquad (V) \qquad (x\to\infty),$$

where V is the Valiron method of summability (cf. Hardy (1949), §§9.10, 9.16). Statement (ii), of 'moving-average' type, is known to be equivalent to

$$I_A(.)\to c \quad (R(e^{\sqrt{n}},1))$$

where $R(e^{\sqrt{n}},1)$ is a Riesz mean of order 1 (cf. Hardy (1949), §§4.16, 5.16); for the equivalence, see Bingham (1981), Bingham and Goldie (1988). But for bounded functions, V and $R(e^{\sqrt{n}},1)$ are known to be equivalent (Bingham and Tenenbaum (1986)).

The density condition (ii) is strictly stronger than the Cesàro density condition in Theorem 1; see Bingham (1981), §1. The Riesz and Valiron methods above are closely linked to the Euler and Borel methods; see §3 below, and for background, Bingham (1984a), (1984c).

Somewhat more classical are the corresponding results for standard (driftless) Brownian motion. Recall the arc-sine law – the law on [0,1] with density $1/(\pi x^{\frac{1}{2}}(1-x)^{\frac{1}{2}})$. The next result is the Brownian analogue of results of Davydov and Ibragimov (1971), Davydov (1973), (1974); cf. Bingham and Goldie (1982), Th.B.

THEOREM 3. *For $A\subset \mathbf{R}^+$ and B standard Brownian motion, the following are equivalent:*

(i) $$\frac{1}{t}|A\cap[0,t]|\to c,$$

(ii) $$P(B_t\in A)\to c,$$

(iii) $$\frac{1}{t}\int_0^t I(B_u\in A)du \quad \textit{converges in law.}$$

and then the limit law is that of $c\xi$ where ξ is arc-sine.

The special case $A=\mathbf{R}^+, c=1$ is Lévy's arc-sine law (Lévy (1939)). For a modern proof of this classical result, see Williams (1979), III.38.10, or Rogers and Williams (1987) VI.53; further references are Kac (1951), Itô and McKean (1965), p.57, Williams (1969), Takács (1981), Pitman and Yor (1986), Karatzas and Shreve (1988), p.273 and p.422.

COROLLARY. (Lévy's arc-sine law).

$\frac{1}{t}\int_0^t I(B_u \geq 0)du$ has the arc-sine law for each $t > 0$.

Proof. By Brownian scaling, the law is the same for each t, and so coincides with the limit law as $t \to \infty$, which is arc-sine by the theorem.

Proof of Theorem 3. We now deduce the theorem from the corollary (showing the equivalence of the two results). We present a streamlined proof in the spirit of the proof of Theorem 1. Note that (i) is

$$\text{(i}'\text{)} \qquad \frac{1}{t}\int_0^\infty g_1(x/t)I_A(x)dx \to c \quad (t \to \infty)$$

with $g_1 := I_{[0,1]}$, while (ii) is

$$\text{(ii}'\text{)} \qquad \frac{1}{t}\int_0^\infty g_2(x/t)I_A(x)dx \to c \quad (t \to \infty)$$

with $g_2(x) := \exp\{-\frac{1}{2}x^2\}/\sqrt{(2\pi)}$. Now g_1, g_2 have Mellin transforms

$$\hat{g}_1(s) := \int_0^\infty g_1(x)x^{is}dx = \int_0^1 x^{is}dx = 1/(1+is),$$

$$\hat{g}_2(s) := \frac{1}{\sqrt{2\pi}}\int_0^\infty e^{-\frac{1}{2}x^2}x^{is}dx = \tfrac{1}{2}.2^{\frac{1}{2}is}\Gamma(\tfrac{1}{2}+\tfrac{1}{2}is)/\sqrt{\pi}.$$

Both are non-zero for all real s, so both g_1 and g_2 may be used as Wiener kernels in the Mellin form of Wiener's Tauberian theorem (Hardy (1949), Th.232) since $f := I_A$ is bounded.

(i) $\Rightarrow$ (ii): Use Wiener's theorem as above (Davydov and Ibragimov (1971)).

(i) $\Rightarrow$ (iii): The measures μ_t defined by

$$\mu_t(x) := \frac{1}{t}\int_0^{tx} I_A(y)dy$$

converge weakly to c times Lebesgue measure on $\mathbf{R}^+$ as $t \to \infty$. Also, by Brownian scaling, $L(t,x) = cL(t/c^2, x/c)$ in law. So

$$\begin{aligned}
\frac{1}{t}\int_0^t I_A(B_u)du &= \frac{1}{t}\int_A dyL(t,y) \\
&= \frac{1}{t}\int_A dy.\sqrt{t}L(1, y/\sqrt{t}) \quad \text{in law} \\
&= \int I_A(v\sqrt{t})L(1,v)dv \\
&= \int L(1,v)d\mu_{\sqrt{t}}(v) \\
&\to c\int_0^\infty L(1,v)dv
\end{aligned}$$

(by compact support of $L(1,v)$)

$$= c\int_0^1 I_{\mathbf{R}^+}(B_u)du$$
$$= c\xi,$$

with ξ arc-sine by Lévy's result.

(iii) $\Rightarrow$ (i): Taking expectations,

$$\int EL(1,v)d\mu_{\sqrt{t}}(v) \to \tfrac{1}{2}c.$$

Now

$$EL(1,v) = \int_0^1 \frac{e^{-\frac{1}{2}v^2/t}}{\sqrt{2\pi}}\frac{dt}{\sqrt{t}} = f(v), \text{ say,}$$

where

$$\begin{aligned}\int_0^\infty f(v)v^{is}dv &= \int_0^1 t^{-\frac{1}{2}}dt\int_0^\infty e^{-\frac{1}{2}v^2/t}v^{is}dv/\sqrt{2\pi}\\ &= \int_0^1 t^{\frac{1}{2}is}dt\int_0^\infty e^{-\frac{1}{2}u^2}u^{is}du/\sqrt{2\pi}\\ &= (1+\tfrac{1}{2}is)^{-1}\hat{g}_2(s),\end{aligned}$$

which is non-zero for real s as above. Thus f is a Wiener kernel, and (i) follows by Wiener's Tauberian theorem as above.

To obtain strong-law behaviour as in Theorem 1, one needs to coarsen the Cesàro averaging, rather than refining it as in Theorem 2. The appropriate summability method is the logarithmic one (or Riesz mean $R(\log n, 1)$; Hardy (1949), Th.37). Logarithmic averages were introduced in probability theory by Lévy (1937), 270 (cf. Chung and Erdős (1951), Th.6, Erdős and Hunt (1953), Th.4); the result below may thus be dubbed 'Lévy's strong arc-sine law'. For extensions, see Révész's contribution to this volume.

THEOREM 4. $\frac{1}{\log t}\int_1^t I(B_u \geq 0)du/u \to \frac{1}{2}$ a.s. $(t\to\infty)$.

First Proof. Writing $u = e^v$ and replacing t by e^t, we have to show

$$\frac{1}{t}\int_0^t I(B(e^v)\geq 0)dv \to \tfrac{1}{2} \quad \text{a.s.} \quad (t\to\infty).$$

Now $Y(t) := e^{-\frac{1}{2}t}B(e^t)$ is an Ornstein-Uhlenbeck process (see e.g. Karlin and Taylor (1981), 380), so we have to show

$$\frac{1}{t}\int_0^t I_{[0,\infty)}(Y_v)dv \to \tfrac{1}{2} \quad \text{a.s.} \quad (t\to\infty).$$

Now the speed measure m of an Ornstein-Uhlenbeck process is finite, and so may be scaled to a probability measure π, which is Gaussian with mean zero. This follows from the stochastic differential equation for the Ornstein-Uhlenbeck process: see Rogers and Williams (1987), V.5.2(ii), V.52.1-2. The result now follows from the ergodic theorem for diffusions,

$$\frac{1}{t}\int_0^t f(Y_u)du \to \int f(x)d\pi(x) \quad \text{a.s.} \quad (t \to \infty),$$

with $f = I_{[0,\infty)}$ and π Gaussian, mean 0 (Rogers and Williams (1987), V.53.5).

Lévy's strong arc-sine law was rediscovered independently (on an equivalent formulation) by Brosamler (1973), Th.1. Use of the Ornstein-Uhlenbeck process in this context may also be found on Brosamler (1986), 314, (1988), 563-4. We thank Michael Lacey for these observations.

Second Proof. This follows from the pathwise central limit theorem, again taking $f = I_{[0,\infty)}$ and using symmetry of a mean-zero Gaussian measure. See Brosamler (1988), Th.1.6; cf. Schatte (1988), Lacey and Philipp (1989+), Fisher (1990+).

The relationship between the three summability methods used in this section may be expressed by

$$R(e^{\sqrt{n}}, 1) \subset R(n, 1) \subset R(\log n, 1).$$

The general result, comparing $R(\lambda_n, k)$ for different λ_n and the same k, is the first consistency theorem for Riesz means; see e.g. Chandrasekharan and Minakshisundaram (1952), Ch.1.

§2. Cesàro and Riesz means

We turn now to more traditional links between summability methods and strong laws. Let $X, X_1, X_2, \ldots$ be independent and identically distributed (iid) random variables. The classical Kolmogorov strong law

$$E|X| < \infty \text{ and } EX = \mu \Leftrightarrow \frac{1}{n}\sum_1^n X_k \to \mu \text{ a.s.}$$

may be rephrased as

$$X \in L_1 \text{ and } EX = \mu \Leftrightarrow X_n \to \mu \text{ a.s. } (C),$$

where $C(= C_1)$ is the Cesàro method of summability. There is a Cesàro method C_α for every positive α (Hardy (1949), V-VII); it was shown by Lai (1974a) that C may be replaced here by C_α for any $\alpha \geq 1$, or by the Abel method A. There are similar versions of the law of the iterated logarithm (Gaposhkin (1965), Lai (1974a)).

For $0 < \alpha < 1$ the situation is different: a.s. C_α-convergence is tied to membership of $L_{1/\alpha}$, not to L_1: for $p \geq 1$,

$$X \in L_p \text{ and } EX = \mu \Leftrightarrow X_n \to \mu \text{ a.s. } (C_{1/p})$$

(Déniel and Derriennic (1988)).

One may improve the forward implication here (which is the harder and more important) by replacing $C_{1/p}$ by a more stringent summability method. It turns out that such a method is provided by the Riesz mean $R_p := R(\exp \int_1^n dx/x^{1/p}, 1) : R_p \subset C_{1/p}$. For $p = 1$, $R_1 = C_1$, but the inclusion is strict for $p > 1$; for details see Bingham (1989).

The Riesz formulation also extends to moments more general than powers. For suitable functions ϕ, Riesz means $R_\phi := R(\exp \int_1^n dx/\phi(x), 1)$ may be linked similarly with membership of a class of Orlicz type, $L_\phi := \{X : E\phi^{\leftarrow}(|X|) < \infty\}$:

$$X \in L_\phi \text{ and } EX = \mu \Leftrightarrow X_n \to \mu \text{ a.s. } (R_\phi).$$

Also, R_ϕ may be written as a summability method of moving-average (or 'delayed-average') type (Bingham and Goldie (1988); Chow (1973); Lai (1974b)): Riesz convergence here is

$$\frac{1}{u\phi(x)} \sum_{x \leq n < x + u\phi(x)} X_n \to \mu \text{ a.s. } \forall u > 0.$$

This moving-average formulation allows one to use results of LIL type by de Acosta and Kuelbs (1983). These authors also consider the Banach-valued case. Further, they give detailed results for the case of slow growth of ϕ – $\phi(x) = c \log x$, or $o(\log x)$ – when strong laws of the above type break down. They are replaced by results of Erdős-Rényi type, where one obtains, instead of the a.s. limit μ above ('a.s. invariance principle'), an a.s. limit superior, $\alpha = \alpha(u)$, which as u varies completely determines the law of X ('a.s. non-invariance principle'). For background on the invariance/non-invariance dichotomy, see Deheuvels and Steinebach (1987).

§3. Euler, Borel and related methods

We recall the classical summability methods of Euler ($E(\lambda), \lambda > 0$) and Borel (B); see Hardy (1949), VIII, IX. These are closely related; methods of Euler-Borel type are perhaps the most important classical summability methods after those of Cesàro-Abel type. They possess an analogue of the above law of large numbers (Chow (1973)) and law of the iterated logarithm (Lai (1974a)), displayed as the Euler and Borel cases of Theorems 5 and 6 below.

In the proofs of these results for $E(\lambda)$ and B, the most important feature of the Euler weights $\binom{n}{k}\lambda^k/(1+\lambda)^n$ and the Borel weights $e^{-x}x^k/k!$ is that they arise (for $x = n$) as n-fold convolutions of the binomial and Poisson distributions respectively, allowing use of the central limit theorem in some form. One may seek to generalise this, and consider weighted sums $\sum a_{nk}X_k$, where the matrix $A = (a_{nk})$ is of convolution type:

$$a_{nk} = P(S_n = k).$$

Here $S_n = \sum_1^n \xi_k$, with the ξ_n independent, $\mathbb{Z}$-valued random variables. There are two important cases:

(a) ξ_n identically distributed (with mean m and variance d^2, say). Then (S_n) is a random walk, S_n has mean nm and variance nd^2, and A is called a summability method of random-walk type (Bingham (1984b), (1984c)).

(b) ξ_n $\{0,1\}$-valued (Bernoulli): $P(\xi_n = 1) = p_n$, say, $P(\xi_n = 0) = q_n := 1 - p_n$. Then the Bernoulli sum S_n has mean $\mu_n := \sum_1^n p_k$, variance $\sigma_n^2 := \sum_1^n p_k(1-p_k)$. Writing $p_n = 1/(1+d_n)(d_n \geq 0)$, one then has

$$\prod_{j=1}^{n} \left(\frac{x + d_j}{1 + d_j} \right) \equiv \sum_{k=0}^{n} a_{nk} x^k.$$

The method $A = (a_{nk})$ is the Jakimovski method $[F, d_n]$ (Jakimovski (1959); Zeller and Beekmann (1970); Ergänzungen, §70). The motivating examples are:

(i) $d_n = 1/\lambda$, the Euler method $E(\lambda)$ above

(ii) $d_n = (n-1)/\lambda$, the Karamata-Stirling method $KS(\lambda)$ (Karamata (1935); Bingham (1988)).

THEOREM 5. *The following are equivalent:*

(i) $\text{var}X < \infty$, $EX = \mu$

(ii) $X_n \to \mu$ a.s. $(E(\lambda)$, or $B)$

(iii) $X_n \to \mu$ a.s. (A), *for A a random-walk method*

(iv) $X_n \to \mu$ a.s. $(KS(\lambda))$, *for some (all) $\lambda > 0$*

(v) $X_n \to \mu$ a.s. $[F, d_n]$, *for $d_n \geq \varepsilon > 0$ for some ε and large n.*

THEOREM 6. *The following are equivalent:*

(i) $EX = 0$, $\text{var}X = \sigma^2$, $E(|X|^4/\log^2|X|) < \infty$

(ii) $\limsup_{x\to\infty} \frac{(4\pi x)^{\frac{1}{4}}}{\log^{\frac{1}{2}} x} \left|\sum_0^\infty e^{-x}\frac{x^k}{k!}X_k\right| = \sigma$ a.s.

(iii) $\limsup_{n\to\infty} \frac{(4\pi n)^{\frac{1}{4}}}{\log^{\frac{1}{2}} n} \left|\sum_0^n \binom{n}{k}\lambda^k X_k/(1+\lambda)^n\right| = \sigma(1+\lambda)^{\frac{1}{4}}$ a.s.

(iv) $\limsup_{n\to\infty} \frac{(4\pi n)^{\frac{1}{4}}}{\log^{\frac{1}{2}} n} \left|\sum a_{nk}X_k\right| = \sigma a^{\frac{1}{4}}$ a.s.

where $A = (a_{nk})$ is a random-walk method with mean-variance ratio $a := m/d^2$,

(v) $\limsup_{n\to\infty} \frac{(4\pi\lambda \log n)^{\frac{1}{4}}}{\log\log^{\frac{1}{2}} n} \left|\sum_0^n a_{nk}X_k\right| = \sigma$ a.s., with $A = KS(\lambda)$

(vi) $\limsup_{n\to\infty} \frac{(4\pi\mu_n)^{\frac{1}{4}}}{\log^{\frac{1}{2}} \mu_n} \left|\sum_0^n a_{nk}X_k\right| = \sigma$ a.s.

with $A = [F, d_n]$ a Jakimovski method with $d_n \to \infty$.

Here the Euler and Borel parts are due to Chow (1973) and Lai (1974a) respectively; the random-walk parts are in Bingham and Maejima (1985); the Jakimovski and

Karamata-Stirling parts are in Bingham and Stadtmüller (1990). The proofs proceed by using normal approximation on the weights a_{nk}, specifically Petrov's local limit theorem (Petrov (1975), VII.3, Th.16) to reduce to the case

$$a_{nk} = \frac{1}{\sigma(2\pi n)^{\frac{1}{2}}} \exp\{-\tfrac{1}{2}(k - n\mu)^2/n\sigma^2\}$$

(or analogue in the Bernoulli, non-identically distributed case). This reduces to the Valiron summability method (Bingham (1984c); cf. the proof of Theorem 2), and one argues as in Lai (1974a),(16). We note in passing that Poisson, rather than normal, approximation is also possible (Bingham and Stadtmüller (1990),§4.2). This involves the Chen-Stein method, which has been studied extensively recently; see for instance Stein (1986); Barbour (1987); Arratia, Goldstein and Gordon (1989).

We note the the $KS(\lambda)$ methods have numerous probabilistic uses, in contexts such as random permutations, records, and greatest convex minorants; for details and references, see Bingham (1988), §3.2. Recent applications include work of Hansen (1987), (1990) on random mappings and the Ewens sampling formula of mathematical genetics.

§4. Complements

1. *Bernstein polynomials.* The classical proof of the Weierstrass approximation theorem (due to S.N. Bernstein in 1912),

$$f(x) = \lim_{n\to\infty} \sum_{0}^{n} f(k/n) \binom{n}{k} x^k (1-x)^{n-k}, \quad f \in C[0,1],$$

has led to many results linking laws of large numbers with summability methods (here the Euler, but others also); for background see Lorentz (1953); Feller (1971), VII; Goldstein (1975), (1976).

2. *Density estimation.* The Bernstein approximation theorem provides one route into the important subject of density estimation, specifically, estimators of smoothed histogram type. For details and references, see Gawronski (1985).

3. *Non-parametric regression.* Asymptotics of matrix transforms $\sum a_{nk}X_k$ have applications to non-parametric estimation of regression curves. For details and references, see Stadtmüller (1984); Lai and Weh (1982).

4. *Time series.* Similarly, the a.s. behaviour of sums $\sum a_{nk}X_k$ has applications to time-series models; see Lai and Weh (1982).

REFERENCES

A. de ACOSTA and J. KUELBS (1983): Limit theorems for moving averages of random vectors. *Z. Wahrschein.* **64**, 67-123.

R. ARRATIA, L. GOLDSTEIN and L. GORDON (1989): Two moments suffice for Poisson approximation: the Chen-Stein method. *Ann. Probab.* **17**, 9-25.

A.D. BARBOUR (1987): Asymptotic expansions in the Poisson limit theorem. *Ann. Probab.* **15**, 748-766.

H.C.P. BERBEE (1987): Convergence rates in the strong law for bounded mixing sequences. *Probab. Th. Rel. Fields* **74**, 255-276.

N.H. BINGHAM (1975): Fluctuation theory in continuous time. *Adv. Appl. Probab.* **7**, 705-766.

N.H. BINGHAM (1981): Tauberian theorems and the central limit theorem. *Ann. Probab.* **9**, 221-231.

N.H. BINGHAM (1984a): On Euler and Borel summability. *J. London Math. Soc.* (2) **29**, 141-146.

N.H. BINGHAM (1984b): Tauberian theorems for summability methods of random-walk type. *J. London Math. Soc.* (2) **30**, 281-287.

N.H. BINGHAM (1984c): On Valiron and circle convergence. *Math. Z.* **186**, 273-286.

N.H. BINGHAM (1988): Tauberian theorems for Jakimovski and Karamata-Stirling methods. *Mathematika* **35**, 216-224.

N.H. BINGHAM (1989): Moving averages. *Almost Everywhere Convergence I* (ed. G.A. Edgar and L. Sucheston), Academic Press, 131-144.

N.H. BINGHAM and C.M. GOLDIE (1982): Probabilistic and deterministic averaging. *Trans. Amer. Math. Soc.* **269**, 453-480.

N.H. BINGHAM and C.M. GOLDIE (1988): Riesz means and self-neglecting functions. *Math. Z.* **199**, 443-454.

N.H. BINGHAM and M. MAEJIMA (1985): Summability methods and almost-sure convergence. *Z. Wahrschein.* **68**, 383-392.

N.H. BINGHAM and U. STADTMÜLLER (1990): Jakimovski methods and almost-sure convergence. *Disorder in Physical Systems* (*J.M. Hamersley Festschrift*, ed. G.R. Grimmett and D.J.A. Welsh), Oxford University Press, 5-18.

N.H. BINGHAM and G. TENENBAUM (1986): Riesz and Valiron means and fractional moments. *Math. Proc. Cambridge Phil. Soc.* **99**, 143-149.

G.A. BROSAMLER (1973): The asymptotic behaviour of certain additive functionals of Brownian motion. *Invent. Math.* **20**, 87-96.

G.A. BROSAMLER (1986): Brownian occupation times on compact manifolds. *Seminar on Stochastic Processes 1985*, 290-322, Birkhäuser.

G.A. BROSAMLER (1988): An almost-everywhere central limit theorem. *Math. Proc. Cambridge Phil. Soc.* **104**, 561-574.

K. CHANDRASEKHARAN and S. MINAKSHISUNDARAM (1952): *Typical Means*, Oxford University Press.

Y.-S. CHOW (1973): Delayed sums and Borel summability of independent, identically distributed random variables. *Bull. Inst. Math. Acad. Sinica* **1**, 207-220.

K.-L. CHUNG and P. ERDŐS (1951): Probability limit theorems assuming only the first moment, I. *Memoirs Amer. Math. Soc.* **6** (four papers on probability).

Yu.A. DAVYDOV (1973): Limit theorems for functionals of processes with independent increments. *Th. Probab. Appl.* **18**, 431-441.

Yu.A. DAVYDOV (1974): Sur une classe de fonctionelles des processus stables et marches aléatoires. *Ann. Inst. H. Poincaré (B)* **10**, 1-29.

Yu.A. DAVYDOV and I.A. IBRAGIMOV (1971): On asymptotic behaviour of some functionals of processes with independent increments. *Th. Probab. Appl.* **16**, 162-167.

P. DEHEUVELS and J. STEINEBACH (1987): Exact convergence rates in strong approximation laws for large increments of partial sums. *Probab. Th. Rel. Fields* **76**, 369-393.

Y. DÉNIEL and Y. DERRIENNIC (1988): Sur la convergence presque sûre, au sens de Cesàro d'ordre α, $0 < \alpha < 1$, des variables aléatoires indépendantes et identiquement distribuées. *Probab. Th. Rel. Fields* **79**, 629-636.

P. ERDŐS and G.A. HUNT (1953): Changes of signs of sums of random variables. *Pacific J. Math.* **3**, 678-687.

W. FELLER (1971): *An Introduction to Probability Theory and its Applications, Vol.II*, 2nd ed., Wiley.

A. FISHER (1990+): A pathwise central limit theorem for random walks. *Ann. Probab.*

V.F. GAPOSHKIN (1965): The law of the iterated logarithm for Cesàro's and Abel's methods of summation. *Th. Probab. Appl.* **10**, 411-420.

W. GAWRONSKI (1985): Strong laws for density estimators of Bernstein type. *Period. Math. Hungar.* **16**, 23-43.

J.A. GOLDSTEIN (1975): Some applications of the law of large numbers. *Bol. Soc. Bras. Mat.* **6**, 25-38.

J.A. GOLDSTEIN (1976): Semigroup-theoretic proofs of the central limit theorem and other limit theorems of analysis. *Semigroup Forum* **12**, 189-206.

J.C. HANSEN (1987): A functional central limit theorem for random mappings. *Ann. Probab.* **17**, 317-332.

J.C. HANSEN (1990): A functional central limit theorem for Ewens' sampling formula. *J. Appl. Probab.* **27**, 28-43.

G.H. HARDY (1949): *Divergent Series.* Oxford University Press.

G. HÖGNÄS and A. MUKHERJEA (1984): On the limit of the average of values of a function at random points. *Lecture Notes in Math.* **1064**, Springer, 204-218 (Probab. Groups VII).

K. ITÔ and H.P. McKEAN (1965): *Diffusion Processes and Their Sample Paths.* Springer, Berlin.

A. JAKIMOVSKI (1959): A generalisation of the Lototsky method. *Michigan Math. J.* **6**, 277-296.

M. KAC (1951): On some connections between probability theory and differential and integral equations. *Proc. 2nd Berkeley Symp. Math. Statists. Prob.*, University of California Press, Berkeley, 189-215.

J. KARAMATA (1935): Théoremès sur la sommabilité exponentielle et d'autres sommabilités s'y rattachant. *Mathematica (Cluj)* **9**, 164-178.

I. KARATZAS and S.E. SHREVE (1988): *Brownian Motion and Stochastic Calculus.* Springer, Berlin.

S. KARLIN and H.L. TAYLOR (1981): *A Second Course in Stochastic Processes.* Academic Press.

W.S. KENDALL and M. WESTCOTT (1987): One-dimensional classical scattering processes and the diffusion limit. *Adv. Appl. Probab.* **19**, 81-105.

M. LACEY and W. PHILIPP (1990): A note on the almost-sure central limit theorem. *Statist. Probab. Letters* **9**, 201-205.

T.-L. LAI (1974a): Summability methods for independent, identically distributed random variables. *Proc. Amer. Math. Soc.* **45**, 253-261.

T.-L. LAI (1974b): Limit theorems for delayed sums. *Ann. Probab.* **2**, 432-440.

T.-L. LAI and C.Z. WEI (1982): A law of the iterated logarithm for double arrays of independent random variables, with applications to regression and time-series models. *Ann. Probab.* **10**, 320-335.

P. LÉVY (1937): *Théorie de l'Addition des Variables Aléatoires.* Gauthier-Villars, Paris.

P. LÉVY (1939): Sur certaines processus stochastiques homogènes. *Compositio Math.* **7**, 283-339.

G.G. LORENTZ (1953): *Bernstein Polynomials.* University of Toronto Press.

I. MEILIJSON (1973): The average of the values of a function at random points. *Israel J. Math.* **15**, 193-203.

V.V. PETROV (1975): *Sums of Independent Random Variables.* Springer.

J.W. PITMAN and M. YOR (1986): Asymptotic properties of planar Brownian motion. *Ann. Prob.* **14**, 733-778.

L.C.G. ROGERS and D. WILLIAMS (1987): ***Diffusions, Markov Processes and Martingales, Volume 2: Itô Calculus.*** Wiley.

P. SCHATTE (1988): On strong versions of the central limit theorem. *Math. Nachrichten* **137**, 249-256.

U. STADTMÜLLER (1984): A note on the law of the iterated logarithm for weighted sums of random variables. *Ann. Statist.* **12**, 35-44.

A.J. STAM (1968): Laws of large numbers for functionals of random walks with positive drift. *Compositio Math.* **19**, 299-333.

C. STEIN (1986): Approximate computation of expectations. *Inst. Math. Statist. Lecture Notes* **7**.

L. TAKACS (1981): The arc-sine law of P. Lévy. *Contributions to Probability* (*Lukács Festschrift*, ed. Gani and Rohatgi), Academic Press, New York, 49-63.

D. WILLIAMS (1969): Markov properties of Brownian local time. *Bull. Amer. Math. Soc.* **75**, 1035-1036.

D. WILLIAMS (1979): *Diffusions, Markov Processes and Martingales, Volume 1: Foundations.* Wiley.

K. ZELLER and W. BEEKMANN (1970): *Theorie der Limitierungsverfahren.* Springer.

Department of Mathematics
Royal Holloway and Bedford New College
Egham Hill, Egham
Surrey TW20 0EX
England

School of Mathematical Sciences
Queen Mary & Westfield College
Mile End Road
London E1 4NS
England

Concerning Induced Operators and Alternating Sequences

R. E. Bradley
Department of Mathematics
Northwestern University
Evanston, IL 60208

July 26, 1990

Abstract

When a positive L_p-contraction T has a semi-invariant function, then it is related in a natural way to a contraction of L_r, where p and r are arbitrary indices in the range $(1,\infty)$. General theorems concerning the pointwise convergence of alternating sequences have been obtained using these induced operators (see [AB,B]). In this note, we study the norm convergence of such sequences, and investigate the existence and uniqueness of induced operators when T is an L_1-contraction.

1 Introduction

Suppose that $p \in (1,\infty)$ and that T is a bounded linear operator on $L_p(\mathbf{X})$, where $\mathbf{X} = (X, \mathcal{F}, \mu)$ is a σ-finite measure space. Let L_p^+ denote the subset of $L_p(\mathbf{X})$ consisting of functions whose range is the non-negative reals (N.B. we are considering complex-valued L_p spaces). We say that T is **positive** if $TL_p^+ \subseteq L_p^+$.

Suppose that $u \in L_p^+$ and let $v = Tu$. We say that u is **semi-invariant** for T if u and v both have full support and $\|v\|_p = \|u\|_p$.

ISBN 0-12-085520-8

If T is a positive L_p-contraction and u is semi-invariant for T, then for any $r \in (1, \infty)$, the equation

$$T_r f = v^{\frac{p}{r}-1} T\left(u^{1-\frac{p}{r}} f\right), \tag{1}$$

for $f \in L_r(\mathbf{X})$, defines a contraction of $L_r(\mathbf{X})$ called the L_r-operator **induced** by T. This operator is independent of the choice of u. Details of the existence and independence proofs may be found in [AB].

An operator T is called **bistochastic** if $T\mathbf{1} = T^*\mathbf{1} = \mathbf{1}$. If T is a bistochastic operator on the L_p space of a finite-measure space, then $\mathbf{1}$ is clearly semi-invariant for T. It has long been known that a positive bistochastic operator is an $L_1 - L_\infty$ contraction; that is, a contraction of L_p for every $p \in [1, \infty]$.

Positive L_p-isometries provide other examples of contractions with semi-invariant functions. Such operators are induced by non-singular point transformations. It is routine to verify that if T is an L_p-isometry induced by a non-singular point transformation τ, then T_r is the L_r-isometry induced by τ.

In [AB], the following theorem was proved; we note that if u is semi-invariant for T, then v^{p-1} is semi-invariant for T^*.

Theorem 1 *Suppose $p, r \in (1, \infty)$. Let $\{T_n\}_{n=1}^{\infty}$ be a sequence of positive L_p-contractions with semi-invariant functions defined over a σ-finite Lebesgue space* $\mathbf{X}$. *Then*

$$(T_1^*)_r \cdots (T_n^*)_r \left[T_n \cdots T_1 f\right]^{\frac{p}{r}}$$

converges a.e. for every $f \in L_p(\mathbf{X})$.

Here, as elsewhere in this paper, $f^{\frac{p}{r}}$ is shorthand for $\psi_{p,r} f$, where $\psi_{p,r} : L_p \to L_r$ is the embedding defined by

$$\left[\psi_{p,r} f\right](x) = \mathrm{sgn} f(x) \left|f(x)\right|^{\frac{p}{r}},$$

for $f \in L_p$, where $\mathrm{sgn}(z)$ is the complex number of unit modulus having the same argument as z. We note that if $f \in L_p$, then

$$\|f^{\frac{p}{r}}\|_r = \|f\|_p^{\frac{p}{r}}.$$

$(X, \mathcal{F}, \mu)$ is a Lebesgue space if X is a separable metric space and $\mathcal{F}$ is the Borel σ-algebra.

Theorem 1 with $r = p$ generalizes Rota's alternating sequence theorem (see [R]), which asserts the pointwise convergence of the sequence

$$T_1^* \cdots T_n^* T_n \cdots T_1 f$$

when the operators are bistochastic.

Theorem 1 also generalizes the theorem of Akcoglu and Sucheston, which asserts the pointwise convergence of the sequence

$$T_1^* \cdots T_n^* [T_n \cdots T_1 f]^{p-1}$$

when the operators are simply positive contractions; see [AS2]. In this case, we have $r = q = p(p-1)^{-1}$ (the adjoint index) in theorem 1 and so the additional hypothesis of semi-invariant function is not needed. Akcoglu and Sucheston also proved the L_p convergence of this sequence in [AS1], which does not require the additional hypothesis of positivity.

In section 2 of this paper, we prove that under the hypotheses of theorem 1, the alternating sequence

$$(T_1^*)_r \cdots (T_n^*)_r [T_n \cdots T_1 f]^{\frac{p}{r}}$$

converges in L_p as well as pointwise. Unlike the result in [AS1], we need the positivity hypothesis, as it is necessary for the unambiguous definition of the induced operator.

In section 3 we investigate the possibility of defining an induced operator for an L_1-contraction. It turns out that the above scheme does yield an L_r contraction T_r when T is an L_1-contraction, but the operator depends on the choice of semi-invariant function.

2 L_p Convergence of Alternating Sequences

For all the results in this section, $p, r \in (1, \infty)$. We say that $\{T_n\}_{n=1}^{\infty}$ is an **inducible** sequence if **X** is a σ-finite Lebesgue space and, for each $n \geq 1$,

$T_n : L_p(\mathbf{X}) \to L_p(\mathbf{X})$ is a positive L_p-contraction with a semi-invariant function. For each $n \geq 1$, let

$$\begin{aligned} V_n &= T_n \cdots T_1, \\ U_n &= (T_1^*)_r \cdots (T_n^*)_r . \end{aligned}$$

Let $f_0 = f^{\frac{p}{r}}$ and, for each $n \geq 1$,

$$f_n = U_n \, [V_n f]^{\frac{p}{r}} .$$

Let $M_n(f) = f_n^{\frac{r}{p}}$ for each $n \geq 0$. Clearly, $f_n \in L_r$ and $M_n f \in L_p$. The operator M_n is non-linear but **homogeneous**; i. e. $M_n(\alpha f) = \alpha M_n f$ for any complex number α.

The following is a consequence of a result of Mazur; see [AB].

Lemma 2 (Uniform Continuity of $\psi_{p,r}$) *For every $\varepsilon > 0$ there is an $\eta > 0$, which depends only on ε, p and r, such that*

$$\| f^{\frac{p}{r}} - g^{\frac{p}{r}} \|_r < \varepsilon$$

whenever $\|f\|_p$, $\|g\|_p \leq 1$ and

$$\|f - g\|_p < \eta .$$

The following maximal estimate is the key to this proof of the norm convergence of $M_n f$. Its proof may be found in [AB], where it is called "Estimate B".

Theorem 3 *For every $\eta > 0$ there is a $\delta > 0$, which depends only on η, p and r, such that for any inducible sequence $\{T_n\}_{n=1}^{\infty}$,*

$$\| \sup_{n \geq 0} |f_n - f_0| \, \|_r < \eta \|f_0\|_r$$

whenever $f \in L_p$ is such that

$$\|f\|_p - \lim_{n \geq 0} \|V_n f\|_p < \delta \|f\|_p .$$

Theorem 1 asserts the pointwise convergence of $\{f_n\}_{n=1}^{\infty}$. The following implies the L_p convergence, because of the uniform continuity of $\psi_{p,r}$.

Theorem 4 *If* $\{T_n\}_{n=1}^{\infty}$ *is an inducible sequence and* $f \in L_p$*, then* $M_n f$ *converges in* L_p *as* $n \to \infty$.

Proof: Let $\varepsilon > 0$ be given. Choose $\eta > 0$ corresponding to ε in the uniform continuity of $\psi_{r,p}$; thus

$$\|g - h\|_r < \eta \Rightarrow \|g^{\frac{r}{p}} - h^{\frac{r}{p}}\|_p < \varepsilon$$

whenever $\|g\|_r, \|h\|_r \leq 1$.

By the homogeneity of the M_ns, we may assume without loss of generality that $\|f\|_p \leq 1$. Let $\alpha_n = \|V_n f\|_p$. $\{\alpha_n\}_{n=0}^{\infty}$ is a non-decreasing sequence of non-negative numbers; let $\alpha = \lim \alpha_n$. Identify two cases:

Case 1. $\alpha = 0$. Choose N such that $\alpha_n < \left(\frac{\eta}{2}\right)^{\frac{r}{p}}$ whenever $n \geq N$. Let $n \geq N$. Then

$$\begin{aligned}
\|f_n - f_N\|_r &< \|f_n\|_r + \|f_N\|_r \\
&= \|\,(T_1^*)_r \cdots (T_n^*)_r\,[V_n f]^{\frac{p}{r}}\,\|_r + \|\,(T_1^*)_r \cdots (T_N^*)_r\,[V_N f]^{\frac{p}{r}}\,\|_r \\
&\leq \|\,[V_n f]^{\frac{p}{r}}\,\|_r + \|\,[V_N f]^{\frac{p}{r}}\,\|_r \\
&= \alpha_n^{\frac{p}{r}} + \alpha_N^{\frac{p}{r}} \\
&< \eta,
\end{aligned}$$

since each $(T_k^*)_r$ is an L_r-contraction. Thus,

$$\|M_n f - M_N f\|_p < \varepsilon$$

by the continuity of $\psi_{r,p}$, and so $M_n f$ converges in L_p.

Case 2. $\alpha > 0$. Choose $\delta > 0$ from Estimate B corresponding to η. Choose N such that $\alpha_n - \alpha < \delta\alpha$ whenever $n \geq N$. Let $\tilde{f} = V_N f$ and define $\tilde{T}_n = T_{N+n}$ for $n \geq 1$. Define $\tilde{V}_n$, $\tilde{U}_n$ and $\tilde{f}_n$ in the same manner as V_n, U_n and f_n, but with respect to the sequence $\left\{\tilde{T}_n\right\}_{n=1}^{\infty}$ and the function $\tilde{f}$. We will apply Estimate B to this sequence. Since

$$\begin{aligned}
\|\tilde{f}\|_p - \lim \|\tilde{V}_n \tilde{f}\|_p &= \alpha_N - \alpha \\
&< \delta\alpha \leq \delta\|\tilde{f}\|_p,
\end{aligned}$$

we have

$$\begin{aligned}
\|\tilde{f}_n - \tilde{f}_0\|_r &\leq \|\sup|\tilde{f}_n - \tilde{f}_0|\,\|_r \\
&< \eta\|\tilde{f}_0\|_r \leq \eta.
\end{aligned}$$

Therefore,

$$\begin{aligned}\|f_{N+n} - f_N\|_r &= \left\|(T_1^*)_r \cdots (T_N^*)_r \tilde{f}_n - (T_1^*)_r \cdots (T_N^*)_r \tilde{f}_0\right\|_r \\ &\leq \left\|\tilde{f}_n - \tilde{f}_0\right\|_r < \eta,\end{aligned}$$

since $(T_1^*)_r \cdots (T_N^*)_r$ is a contraction. Thus,

$$\|M_{N+n}f - M_N f\|_p < \varepsilon,$$

whenever $n \geq 0$ and so $M_n f$ converges in L_p. ∎

Remarks: It seems somewhat unsatisfying to use as strong a result as Estimate B in the proof of theorem 4. The proof of Estimate B is lengthly and involves a dilation argument as well as the martingale convergence theorem. Also, Estimate B is enough to give the pointwise convergence of f_n when it is also known that there is maximal inequality for the sequence f_n; historically, L_p convergence arguments in ergodic theory have been obtained more easily than the corresponding pointwise results.

The theorems of Akcoglu & Sucheston involve the sequences f_n and $M_n f$ in the case $r = q$. Here, as already noted, semi-invariant functions are not required in defining f_n and $M_n f$. The pointwise convergence argument for f_n does indeed rest on a proof of Estimate B in this special case (see [AS2]). However, the norm convergence (see [AS1]) is deduced from a lemma which states that $\|f_n - f_0\|_q$ can be made arbitrarily small when f is in the closed unit ball in L_p and $\|f\|_p - \|V_n f\|_p$ is sufficently small. The proof is concise and uses geometric arguments - particularly the convexity of L_p - but it depends upon the fact that

$$\begin{aligned}(f, f_n) &= \left(V_n f, (V_n f)^{p-1}\right) \\ &= \|V_n f\|_p^p\end{aligned}$$

when $r = q$. The analogous result for an arbitrary index r is that

$$\left(f^{\frac{p}{s}}, f_n\right) = \left((T_n)_s \cdots (T_1)_s f^{\frac{p}{s}}, V_n f\right),$$

where $s = r(r-1)^{-1}$. This curiosity follows from the property $(T^*)_s = (T_r)^*$ of induced operators (see [B]) but appears to have no utility in a proof of the norm convergence of $M_n f$.

3 An L_1 counterexample

In [B], it is shown that whenever $p, r \in [1, \infty)$ and T is a positive contraction with semi-invariant function u, then equation 1 defines a positive L_r contraction. Furthermore, the uniqueness proof in [AB] goes through if $r = 1$. Thus, T_r is independent of the choice of u, for any $r \in [1, \infty)$, as long as $p > 1$. In this section, we will see that if T is an L_1-contraction, then the operator T_r generally depends on the choice of u.

Let X be 2-dimensional space and let T be given by a bistochastic matrix with strictly positive entries. For example, let the matrix of T be

$$\begin{bmatrix} \frac{1}{2} & \frac{1}{2} \\ \frac{1}{2} & \frac{1}{2} \end{bmatrix}$$

with respect to the standard basis.

We will find two semi-invariant functions u_1 and u_2 and define, for $i = 1, 2$,

$$S_i f = v_i^{-\frac{1}{2}} T\left(u_i^{\frac{1}{2}} f\right).$$

That is, the L_2-contraction induced by T, corresponding to each u_i.

Let

$$u_1 = \begin{bmatrix} 1 \\ 1 \end{bmatrix}, u_2 = \begin{bmatrix} 2 \\ 1 \end{bmatrix}.$$

On the one hand, $S_1 = T$. On the other hand, S_2 is the operator given by the matrix

$$\begin{bmatrix} \frac{1}{\sqrt{3}} & \frac{1}{\sqrt{6}} \\ \frac{1}{\sqrt{3}} & \frac{1}{\sqrt{6}} \end{bmatrix}.$$

This shows that induced operators depend on the choice of u when $p = 1$. We note that if we had $p > 1$, then u_1 would be the only semi-invariant function up to scalar multiplicity.

References

[AB] M. A. Akcoglu, R. E. Bradley, Alternating sequences and induced operators, *Trans. AMS*, to appear.

[AS1] M. A. Akcoglu, L. Sucheston, An alternating procedure for operators on L_p spaces, *Proc. AMS*, **99** (1987), 555-558.

[AS2] M. A. Akcoglu, L. Sucheston, Pointwise convergence of alternating sequences, *Can. J. Math.*, **40** (1988), 610-632.

[B] R. E. Bradley, On Induced Operators, *Can. J. Math.*, to appear.

[R] G. C. Rota, An "alternierende Verfahren" for general positive operators, *Bull. Amer. Math. Soc.*, **68** (1962), 95-102.

Maximal inequalities and ergodic theorems for Cesàro-α or weighted averages

M. Broise, Y. Déniel and Y. Derriennic

When Birkhoff proved the ergodic theorem, summability theory was fashionable and in full progress. Yet many problems involving the relationship between ergodic theorems and summability methods were considered only recently. The aim of this paper is to discuss some of them.

In part 1) we give a résumé of what is known about the strong law of large numbers and the ergodic theorem for Cesàro-α averages with $\alpha > 0$. In part 2) we give a general form of the maximal ergodic inequality for weighted averages and some of its corollaries. Such a maximal inequality is essential in the proof of the ergodic theorem for Cesàro-α averages with $0 < \alpha < 1$. In part 3) we extend the "precise" form of the maximal inequality, in dimension 1, to Cesàro-α averages and show that they are the only weighted averages for which this is possible. The results of parts 2) and 3) can be viewed as a complement to reference **[5]**.

In Hardy-Littlewood's work on maximal inequality, the equimeasurable rearrangement of a function or a sequence was an essential tool (**[12]**, **[13]**, chap X). In the sequel it will be clear that this notion is still fundamental for the study of weighted averages.

1) The strong law of large numbers and the ergodic theorem for Cesàro-α averages : a résumé.

Let us consider a measure preserving transformation ϑ acting on a probability space $(\Omega, \mathfrak{F}, \mu)$. The Cesàro-$\alpha$ ergodic averages (i.e. (C,α)) for a real measurable function f on Ω are defined by :

$$\mathfrak{T}_n^\alpha f(\omega) = \begin{pmatrix} n+\alpha \\ n \end{pmatrix}^{-1} \sum_{k=0}^{n} \begin{pmatrix} n-k+\alpha-1 \\ n-k \end{pmatrix} f(\vartheta^k \omega).$$

For a measure-preserving semi-flow $(\vartheta_s)_{s>0}$ the (C,α)-averages are defined by :

$$\mathfrak{T}_t^\alpha f(\omega) = \frac{\alpha}{t^\alpha} \int_0^t (t-s)^{\alpha-1} f(\vartheta_s \omega) ds.$$

These two formulae are similar. The second one may be more transparent.

The properties of these classical summability methods are described in **[11]** and **[27]**. When $\alpha = 1$ the (C,α)-averages reduce to the usual arithmetical averages. Since we want to consider only positive weights we shall consider (C,α)-averages only with $\alpha > 0$. A basic result is that (C,α)-convergence implies (C,β)-convergence whenever $\beta > \alpha$. In other words the (C,α)-method is weaker than the (C,β)-method. When $\alpha \to +\infty$ the Abel averages come in :

ISBN 0-12-085520-8

$$\mathcal{A}b_r\, f(\omega) = (1-r) \sum_{k=0}^{\infty} r^k\, f(\vartheta^k \omega) \;;$$

just take $n \sim p\alpha$ for some constant p and put $r = e^{-p}$. Given any $\alpha > 0$, the (C,α) convergence implies the Abel convergence for $r \to 1^-$. When $\alpha \to 0^+$ the M. Riesz harmonic averages come in :

$$\mathcal{H}_n\, f(\omega) = \frac{c_n}{\log\,(n+1)} \sum_{k=0}^{n} \frac{f(\vartheta^k \omega)}{n+1-k}$$

where c_n is a normalizing constant ~ 1. The Riesz harmonic convergence implies the (C,α) convergence for every $\alpha > 0$ ([**11**]).

From classical Tauberian theory it is known that for bounded sequences all (C,α) convergences for $\alpha > 0$ and also Abel convergence are equivalent ; but they are not equivalent to Riesz harmonic convergence. It is also known that for positive sequences all (C,α) convergences for $\alpha \geq 1$ and also Abel convergence are equivalent ([**11**]).

Kolmogorov's strong law of large numbers asserts that, given an independent identically distributed sequence of random variables X_n, the a.s. (C,1)-convergence is equivalent to $E(|X_1|) < \infty$. In this context it is therefore clear that the only interesting questions about a.s. (C,α) convergence with $\alpha > 1$ or Abel convergence is whether such a convergence does happen without finite first moment. For positive variables the answer is negative because of the Tauberian theorem. It is a remarkable result that in general the answer is also negative ([**18**]) : if a.s. Abel convergence holds for an i.i.d. sequence then it has a finite first moment ; a fortiori if a.s. (C,α) convergence holds. The case $\alpha < 1$ seems to have been considered for the first time in [**21**] where the following result can be found : for an i.i.d. sequence the a.s. (C,α)-convergence is equivalent to $E\left(|X_1|^{1/\alpha}\right) < \infty$ when $1/2 < \alpha \leq 1$. The fact that this result holds for any $0 < \alpha \leq 1$ was obtained only recently ([**10**]) ; the case $0 < \alpha < 1/2$ was shown to be an easy by-product of an abstract result of [**7**] ; the case $\alpha = 1/2$ was given a separate specific proof. More recently two different other approaches to this theorem, which do not require the distinction between the cases $\alpha < 1/2$, $\alpha = 1/2$, $1/2 < \alpha$, were found ([**1**], [**14**]).

Let us come now to the ergodic averages. From Tauberian theory it is clear that a.e. (C,α)-convergence with $\alpha \geq 1$ and Abel convergence are all equivalent for a function $f \in L^1$ or $f \geq 0$ μ a.e.. Knowing Birkhoff's ergodic theorem, the only interesting question about a.e. (C,α)-convergence with $\alpha > 1$ or Abel convergence is whether such a convergence does hold for f in a bigger space than L^1. The answer requires first a little discussion about the necessity of the assumption $f \in L^1$ in Birkhoff's theorem itself.

Let us assume, to avoid technicalities, that the system is ergodic and the probability space $(\Omega, \mathcal{F}, \mu)$ is a non atomic Lebesgue space. If $f \geq 0$ and $\int f d\mu = +\infty$, there is a.e. (C,1) and also

Abel convergence to $+\infty$. But there always exist functions f with $\int f^+ d\mu = \int f^- d\mu = +\infty$ such that a.e. (C,1)-convergence to a finite limit holds. Yet no assumption weaker that $f \in L^1$, bearing only on the distribution of f, can imply the a.e. (C,1) convergence to a finite limit. Already the necessity of $E(|X_1|) < \infty$ in Kolmogorov's strong law is a proof of this statement. In a more precise sense, if

$\int f^+ d\mu = \int f^- d\mu = +\infty$ there is a function f', on the same system $(\Omega, \mathfrak{F}, \mu, \vartheta)$ having the same distribution as f, that is $\mu\{f' > y\} = \mu\{f > y\}$ for every $y \in \mathbb{R}$, such that its (C,1) ergodic averages diverge a.e. with limsup $= +\infty$ a.e. and liminf $= -\infty$ a.e. Such a function f' is an equimeasurable rearrangement of f; it is deduced from f via a measure preserving transformation σ of (Ω, μ) : $f' = f \circ \sigma$. In other words if $\int f^+ d\mu = \int f^- d\mu = +\infty$, there is a measure preserving transformation ϑ' that is conjugate to ϑ i.e. $\vartheta' = \sigma \circ \vartheta \circ \sigma^{-1}$, for which the (C,1)-averages of f diverge a.e. ([5] prop. 6).

Strictly speaking the spaces of functions on $(\Omega, \mathfrak{F}, \mu, \vartheta)$ for which a.e. (C,α) or Abel convergence holds, are all distinct. But if we restrict our consideration to spaces which can be defined by conditions bearing only on the distribution of the functions, in other words to spaces invariant under equimeasurable rearrangements, then L^1 is the bigger space for which a.e. (C,1) or (C,α) with $\alpha > 1$ or Abel-convergence holds. This is a consequence of the necessity of the L^1-assumption for Abel convergence already in the i.i.d. case.

For a.e. (C,α) convergence with $\alpha < 1$ the picture is essentially different. In [16] it was shown that a.e. (C,α) convergence holds as soon as $f \in L^p$ with $\alpha p > 1$. In [9] it was precisely shown that this is not true, in general, if $\alpha p = 1$ and $\alpha < 1$; in other words the condition $E\left(|X_1|^{1/\alpha}\right) < \infty$ equivalent to the a.e. (C,α) convergence in the i.i.d. case is not a sufficient condition in the ergodic case, when $0 < \alpha < 1$. The exact assumption on the distribution of the function which yields a.e. (C,α) convergence in this case was found in [5] : it is f in the Lorentz space $L(1/\alpha, 1)$. Given f, the decreasing rearrangement f^* of f is the right continuous decreasing, in the wide sense, function on (0,1) such that $\lambda\{f^* > y\} = \mu\{|f| > y\}$ for every $y \geq 0$, with λ denoting the Lebesgue measure. Then $L(1/\alpha, 1) = \left\{ f \; ; \; \|f\|^*_{1/\alpha,1} = \alpha \int_0^1 f^*(t)\, t^{\alpha} \frac{dt}{t} < \infty \right\}$.

Since $\mu(\Omega) = 1$ we have $L^p \subset L(1/\alpha, 1) \subset L^{1/\alpha}$ for $\alpha p > 1$. The necessity of the $L(1/\alpha, 1)$ assumption for the a.e. (C,α) convergence when $0 < \alpha < 1$, has the same meaning as the necessity of L^1 for Birkhoff's theorem. If $f \notin L(1/\alpha, 1)$ there is an equimeasurable rearrangement f' of f for which the (C,α) averages diverge a.e. A key argument in this study is a maximal inequality for "rearranged" averages. A general version of it will be described in the next section.

To finish with this résumé we consider the Riesz harmonic means. In the i.i.d. case the a.e. Riesz harmonic convergence is equivalent to $E\left(e^{t|X_1|}\right) < \infty$ for every $t \geq 0$ ([7]). In the case of ergodic averages, the biggest, rearrangement invariant, space of functions for which the a.e. Riesz harmonic convergence holds is the space of constants (**[9]**).

2) Maximal inequality and ergodic theorem for weighted averages : a general formulation.

We begin with the description of a general class of weighted averages in a real analysis setting. We shall introduce the ergodic averages only later.

Let φ be a weight on $\mathbb{R}^d$ $(d \geq 1)$, that is a real measurable function such that :

i) $\varphi(x) \geq 0$ λ a.e. (λ is the Lebesgue measure of $\mathbb{R}^d$)

ii) $\int \varphi(x)dx = 1$

iii) the support of φ is a bounded set S.

For any linear isomorphism τ of $\mathbb{R}^d$ we define the average of a real fun ction f on $\mathbb{R}^d$ as :

$$A_\tau^\varphi f(x) = \frac{1}{|\det \tau|} \int_{\mathbb{R}^d} \varphi(\tau^{-1}z)\, f(x+z)\, dz = \int_{\mathbb{R}^d} \varphi(y)\, f(x+\tau y)\, dy$$

The (C,α) averages $\mathcal{C}_t^\alpha f(x)$, introduced in part **1)** are examples of such averages : take $d = 1$, $\varphi(x) = \alpha(1-x)^{\alpha-1}$ for $0 < x < 1$, and τ the dilation of ratio $t > 0$. Another example, with $d > 1$, will be discussed below. Because of the fundamental inequality about rearrangement (**[13]** chap X, n° 378) which reads as follows : $\int_E fg\ dm \leq \int_0^{m(E)} f^*(t)\, g^*(t)\, dt$ we have

$$|\det \tau|\, |A_\tau^\varphi f(x)| = \left| \int_{\tau S} (\varphi \circ \tau^{-1})(z)\, f(x+z) ds \right| \leq \int_0^{|\det\tau|\lambda(S)} (\varphi \circ \tau^{-1})^*(t) \Big(f(x+.)\chi_{\tau S}(.) \Big)^*(t) dt$$

where the $*$ denotes the decreasing rearrangement as recalled in part **1)**. It is easy to see that

$(\varphi \circ \tau^{-1})^*(t) = \varphi^*(t/|\det\tau|)$ so $|A_\tau^\varphi f(x)| \leq |A_\tau^{*\varphi} f(x)|$ with

$$A_\tau^{*\varphi} f(x) = \frac{1}{|\det \tau|} \int_0^{|\det\tau|\lambda(S)} \varphi^*(t/|\det\tau|) \Big(f(x+.)\chi_{\tau S}(.) \Big)^*(t) dt$$

(supremum of the averages with respect to all the weights having the same distribution as φ).

In the sequel, since φ is fixed, we shall omit the superscripts φ in the notations.

For any τ, linear isomorphism of $\mathbb{R}^d$, we put $k(\tau) = \|\tau\| . \|\tau^{-1}\|$ where the norm is the euclidean norm (this choice is unessential). Some authors call this number the "condition number" of τ. The following easy property expresses how this quantity will be used : given a bounded measurable set B in $\mathbb{R}^d$, let $\hat{B}$ be the smallest euclidean ball containing B and $\gamma(B) = \dfrac{\lambda(\hat{B})}{\lambda(B)}$, then we have $\gamma(\tau B) \le k(\tau)^d \gamma(B)$. This property implies at once that the family of sets $(\tau S)_{\tau\in G\ell(d,\mathbb{R}),k(\tau)\le K}$ is "regular" in the sense of Tempel'man ([25] p. 106 ; see also [17] p. 203), since we get $\sup_{k(\tau)\le K} \gamma(\tau S) \le K^d\gamma(S)$.

Then we define the following maximal functions $M_K f(x) = \sup_{k(\tau)\le K} A_\tau f(x)$ and $M_K^* f(x) = \sup_{k(\tau)\le K} A_\tau^* f(x)$. We have $M_K^* f(x) \ge M_K |f|$.

We are going to prove a maximal inequality for M_K and M_K^*, similar to the Hardy-Littlewood inequality. Under a suitable assumption on the shape of φ such a maximal inequality can be deduced directly from the usual one for arithmetic averages, that is averages defined with φ equal to the indicator of S (see [8] chap X or [25] § 7). The essential point in our following statement is that it holds for any weight φ.

Maximal inequality

There is a constant C *depending only on the dimension* d, *the constant* K *and the set* S, *such that for every* $f \ge 0$ *measurable on* $\mathbb{R}^d$ *and every* $y > 0$ *we have*

$$\lambda\{M_K f > y\} \le \lambda\{M_K^* f < y\} \le \frac{C}{y}\int_0^{\lambda\{M_K^* f>y\}} f^*(t)\varphi^*\left(t\lambda(S)/\lambda\{M_K^* f>y\}\right)dt.$$

If $\varphi^*(t) \le Q\alpha t^{\alpha-1}$ *for* $0 < t < \lambda(S)$, *with* $0 < \alpha < 1$ *and a constant* Q, (*for* $t > \lambda(S)$, $\varphi^*(t) = 0$ *by definition*) *there is a constant* C *depending also on* Q *such that*

$$\lambda\{M_K f > y\} \le \lambda\{M_K^* f < y\} \le \left(\frac{C}{y}\|f\|^*_{1/\alpha,1}\right)^{1/\alpha}$$

where $\|f\|^*_{1/\alpha,1} = \alpha\int_0^{+\infty} f^*(t)t^\alpha \frac{dt}{t}$.

Proof : We consider first the case of a bounded, with compact support, f. For fixed $T > 0$ we consider

$$M^*_{K,T} f = \sup_{k(\tau)\le K;|\det\tau|\le T} A^*_\tau f.$$

To abbreviate we denote this function only M. Let $E_y = \{M > y\}$. Because of $k(\tau) \le K$ and $|\det \tau| \le T$, E_y is bounded. For every $x \in E_y$, there is τ with $k(\tau) \le K$ and $|\det \tau| \le T$ such that $A^*_\tau f(x) > y$. The family of sets $(\tau S)_{k(\tau)\le K;|\det \tau|\le T}$ being "regular" the covering lemma can be used ([17], [25]) therefore we can find a finite sequence (x_i, τ_i) such that $A^*_{\tau_i} f(x_i) > y$, the sets $V_i = x_i + (\tau_i S)$ are pairwise disjoint and $\lambda(E_y) \le C\lambda(L)$ where $L = \bigcup_i V_i$ and C is a constant depending only on d and $K^d\gamma(S)$. Using Corollaire 2 of [5] we get

$$y < \frac{1}{\lambda(L)} \sum_i \lambda(V_i)\, A^*_{\tau i} f(x_i) \le \frac{1}{\lambda(L)} \int_0^{\lambda(L)} \varphi^*\!\left(\frac{t\lambda(S)}{\lambda(L)}\right) (f\chi_L)^*(t)\, dt.$$

Since f^* and φ^* are decreasing we obtain

$$y < \int_0^1 f^*(s\lambda(L))\, \varphi^*(s\lambda(S))ds \le \int_0^1 f^*(s\frac{\lambda(E_y)}{C})\, \varphi^*(s\lambda(S))ds = \frac{C}{\lambda(E_y)} \int_0^{\lambda(E_y)/C} f^*(t)\varphi^*\!\left(t\,\frac{C\lambda(S)}{\lambda(E_y)}\right)dt$$

The constant C can be taken ≥ 1 and we get the inequality we were looking for :

$$y < \frac{C}{\lambda(E_y)} \int_0^{\lambda(Ey)} f^*(t)\, \varphi^*\!\left(t\, \frac{\lambda(S)}{\lambda(E_y)}\right) dt.$$

Obvious monotone limits, first in f then in $T \uparrow +\infty$, yield the inequality in the general case. When $\varphi^*(t) \le Q\, \alpha t^{\alpha-1}$ for $0 < t < \lambda(S)$ we have, at once

$$\lambda(E_y)^\alpha < \frac{CQ\lambda(S)^{\alpha-1}}{y} \|f\|^*_{1/\alpha,1} \qquad \blacklozenge$$

Now we consider the ergodic averages

$$\mathcal{A}^\varphi_\tau f(\omega) = \int_{\mathbb{R}^d} \varphi(y) f(\vartheta_{\tau y}\omega)\, dy = \frac{1}{|\det y|} \int_{\mathbb{R}^d} \varphi(\tau^{-1}z)\, f(\vartheta_z\omega)\, dz$$

with respect to a semi flow $(\vartheta_x)_{x\in \mathbb{R}^d_+}$ that is measure preserving on a probability space $(\Omega, \mathcal{F}, \mu)$.

The maximal function is $\mathcal{M}^\varphi_K f(\omega) = \sup_{k(\tau)\le K} \mathcal{A}^\varphi_\tau f(\omega)$.

As before we drop the superscript φ in the notations : the weight is fixed once for all.

Using the transfer principle ([26], [6]) we can deduce from the preceding "real" maximal inequality the following :

Maximal ergodic inequality

For the constant C, *depending only on* d, K *and* S, *given by the preceding "real" maximal inequality, we have, for every measurable* $f \ge 0$ *on* Ω *and every* $y > 0$:

$$\mu\left\{\mathcal{M}_K f > y\right\} \le \frac{C}{y} \int_0^{\mu\{\mathcal{M}_K f>y\}} f^*(t)\, \varphi^*\!\left(t\lambda(S)/_{\mu\{\mathcal{M}_K f>y\}}\right)dt.$$

If $\varphi^*(t) \le Q\alpha t^{\alpha-1}$ *for* $0 < t$ *with* $0 < \alpha < 1$, *and a constant* Q, *there is a constant* C *depending also on* Q *such that* $\mu\{\mathcal{M}_K f > y\} \le \left(\frac{C}{y} \|f\|^*_{1/\alpha,1}\right)^{1/\alpha}$.

Proof : It is similar to the proof given in [5] p. 701, so we skip the details. The transfer principle will be explained again in the next section, for the "precise" form of the maximal inequality. ♦

It is easy to check that, with $k(\tau) \leq K$, when $|\det \tau| \to +\infty$,

$\lim \frac{1}{|\det \tau|} \int |\varphi o \tau^{-1}(x) - \varphi o \tau^{-1}(x-x')|dx = 0$ hence the almost-invariance of the system of averages A_τ, or $\mathcal{Q}_\tau$, with $k(\tau) \leq K$, follows. In other words the averages $(\mathcal{Q}_\tau)$, with $k(\tau) \leq K$ and $|\det \tau| \to +\infty$, are an ergodic net in the sense of [17], chap. 2, acting on the abstract Lorentz spaces :

$\Lambda_\mu(\varphi^*,q) = \left\{ f : \Omega \to \mathbb{R} \ ; \ \|f\|^*_{\varphi^*,q} = \left(\int_0^1 \varphi^*(t\lambda(S))(f^*(s))^q ds\right)^{1/q} < \infty \right\}$ where $q \geq 1$. The mean ergodic theorem holds for these averages, in $\Lambda_\mu(\varphi^*,q)$ with $q > 1$ because this space is reflexive ([20]). It also holds in $\Lambda_\mu(\varphi^*,1)$ because of the inequality $\|f\|^*_{\varphi^*,q} \geq \|f\|^*_{\varphi^*,1}$ where $q > 1$, and the density of $\Lambda_\mu(\varphi^*, q)$ in $\Lambda_\mu(\varphi^*,1)$. Therefore the functions f which are either invariant or of the form $g - g \, o \, \vartheta_x$ with $x \in \mathbb{R}^d_+$ and g bounded, generate a dense subspace in $\Lambda_\mu(\varphi^*,1)$. Then a standard application of the maximal ergodic inequality yields the :

Ergodic theorem for weighted averages, or fractionnal ergodic theorem

For $f \in \Lambda_\mu(\varphi^*,1)$, *the averages* $\mathcal{Q}^\varphi_\tau f(\omega) = \frac{1}{|\det \tau|} \int_{\mathbb{R}^d} \varphi(\tau^{-1}z) f(\vartheta_z \omega)dz$ *are converging* μ a.e. *when* $|\det \tau| \to +\infty$ *with* $k(\tau) \leq K$. *If* $\varphi^*(t) \leq Q\alpha t^{\alpha-1}$ $(t > 0)$ *then this convergence holds for* $f \in L_\mu(1/\alpha,1)$.

We underline again that this statement holds for any weigth φ. Under additional assumptions on the shape of φ the ergodic theorem holds for $f \in L^1_\mu$, which is a bigger space than $\Lambda_\mu(\varphi^*,1)$ ([25], [17]). But in general, without assumption on φ, $\Lambda_\mu(\varphi^*,1)$ is the best possible for the ergodic theorem ([5]).

Using Marcinkiewicz' theorem, strong type maximal inequalities can be proved in $L_\mu(p,q)$ with $p > 1/\alpha$, $q \geq 1$ when $\varphi^*(t) \leq Q\alpha t^{\alpha-1}$ for $t > 0$. We can even obtain the following more precise result, similar to a classical result of Zygmund.

Dominated ergodic theorem

Let the weight φ *be such that* $\varphi^*(t) \leq Q\alpha t^{\alpha-1}$ $(t > 0)$ *with* $0 < \alpha < 1$. *If*
$\alpha \int_0^1 t^{\alpha-1}(f^*(t)\log^+ f^*(t))dt = \|f^* \log^+ f^*\|^*_{1/\alpha,1} < \infty$ *then*

$$\alpha \int_0^1 t^{\alpha-1}(\mathcal{M}_K f)^*(t)dt = \|\mathcal{M}_K f\|^*_{1/\alpha,1} < \infty \quad \text{(i.e. } \mathcal{M}_K f \in L_\mu(1/\alpha,1)).$$

Proof : The proof follows classical lines. First of all integration by parts yields

$$\alpha \int_0^1 g^*(t)t^{\alpha-1}dt = \int_0^\infty \mu\{|g|>t\}^\alpha dt$$

for every real measurable function g on $(\Omega, \mathcal{F}, \mu)$. We may consider $f > 0$. Given $u > 0$, put $g = f\chi_{\{f>\frac{u}{2}\}}$; then $\mathcal{M}_K f \le \mathcal{M}_K g + \frac{u}{2}$ and $\mu\left\{\mathcal{M}_K f > u\right\}^\alpha \le \mu\left\{\mathcal{M}_K g > \frac{u}{2}\right\}^\alpha$.

The maximal ergodic inequality yields $\mu\left\{\mathcal{M}_K g > \frac{u}{2}\right\}^\alpha \le \frac{2C}{u}\int_0^1 g^*(t)\alpha t^{\alpha-1}dt.$

By integration we get $\int_0^\infty \mu\left\{\mathcal{M}_K f > \frac{u}{2}\right\}^\alpha du \le 1 + \int_1^\infty \frac{2C}{u}\left(\int_0^1 \left(f\chi_{\{f>\frac{u}{2}\}}\right)^*(t)\alpha t^{\alpha-1}dt\right)du.$

Since $\left(f\chi_{\{f>\frac{u}{2}\}}\right)^*(t) = 0$ whenever $u \ge 2f^*(t)$, a permutation of the integrations leads to

$$\|\mathcal{M}_K f\|^*_{1/\alpha,1} \le 1 + 2C\int_0^1 \alpha t^{\alpha-1}\left(\int_1^{2f^*(t)} f^*(t)\frac{du}{u}\right)dt$$

$$\le C' + 2C\int_0^1 \alpha t^{\alpha-1}\left(f^*(t)\log^+ f^*(t)\right)dt \qquad \blacklozenge$$

In the proof of the ergodic theorem for multidimensional cocycles [3] weighted averages of the general type studied above, play a fundamental rôle. To end this section we describe this example, where $d > 1$. The weight is

$$\varphi(x) = C\|x - e_1\|^{1-d} \text{ if } \|x - e_1\| < 1$$
$$= 0 \quad \text{otherwise}$$

where $e_1 = (1, 0....0)$ is the first basis vector of $\mathbb{R}^d$ and where C is a normalising constant. The needed averages are $\mathcal{A}_\tau f(\omega) = \frac{1}{|\det \tau|}\int_{\mathbb{R}^d}\varphi(\tau^{-1}z)f(\vartheta_z\omega)dz$ with τ a similitude of $\mathbb{R}^d$, that is a product of a dilation of ratio $t > 0$ and a rotation r. In that case $k(\tau) = 1$, $|\det \tau| = t^d$;we write

$$\mathcal{A}_{t,r} f(\omega) = \frac{1}{t^d}\int_{\mathbb{R}^d}\varphi\left(\frac{r^{-1}z}{t}\right)f(\vartheta_z\omega)dz.$$

Since $\varphi^*(t) \le Qt^{\frac{1}{d}-1}$ the maximal ergodic inequality proved above, has the following application in this example : $\mu\left\{\sup_{t>0;r}\mathcal{A}_{t,r} f > y\right\} \le \left(\frac{C}{y}\|f\|^*_{d,1}\right)^d$ ([3], prop.9)

3) The "precise" form of the maximal inequality

On $\mathbb{R}$ the maximal inequality of Hardy and Littlewood $\lambda\{Mf > y\} \leq \frac{1}{y}\|f\|_1$, where $f \geq 0$ and $Mf(x) = \sup_{t>0} \frac{1}{t}\int_0^t f(x+s)ds$, admits of the more precise following form :

$$\lambda\{Mf > y\} \leq \frac{1}{y}\int_{\{Mf>y\}} f(s)ds$$

(as before λ denotes the Lebesgue measure). In this section we give a similar inequality for the (C,α) averages with $0 < \alpha < 1$ and we show that they are the only weighted averages on $\mathbb{R}$ for which such a "precise" maximal inequality is possible. The (C,α) averages, on $\mathbb{R}$, are the weighted averages with weight $\varphi(x) = \alpha(1-x)^{\alpha-1}$ for $x \in (0,1)$. In the real setting we write

$$C_t^\alpha f(x) = \frac{\alpha}{t^\alpha}\int_0^t (t-s)^{\alpha-1} f(x+s)\, ds.$$

In the ergodic setting we write, as above : $\mathfrak{C}_t^\alpha f(\omega) = \frac{\alpha}{t^\alpha}\int_0^t (t-s)^{\alpha-1} f(\vartheta_s\omega)\, ds.$

We are going to prove the "precise" form of the maximal inequality in the real setting first and then we will adapt the transfer method to prove it in the ergodic setting.

Given a measurable $f \geq 0$ on $\mathbb{R}$ we put $\qquad Mf(x) = \sup_{0<t} C_t^\alpha f(x)$

and $\qquad M_T f(x) = \sup_{0<t<T} C_t^\alpha f(x)$

We consider also $C_t^{*\alpha} f(x) = \frac{\alpha}{t^\alpha}\int_0^t s^{\alpha-1}\left(f(x+.)\chi_{(0,t)}(.)\right)^*(s)ds$

(maximum value of the averages of f on $(x, x+t)$ with respect to all the weights having the same distribution as $\alpha t^{\alpha-1}$), and the corresponding $M^*f(x)$ and $M_T^* f(x)$. Of course we have $C_t^\alpha \leq C_t^{*\alpha}$ and $M \leq M^*$; if $\alpha = 1$ they are equal !

In the classical case, $\alpha = 1$, a key step in the proof of the precise form of the maximal inequality is the "rising sun lemma" of Riesz ([4], [13], [22]). When $0 < \alpha < 1$ we have a similar statement.

Rising sun lemma when $0 < \alpha < 1$.

Let us consider $f \geq 0$, *measurable, bounded, with compact support in* $\mathbb{R}$. *For every* $y > 0$, *the set* $\{M_T f > y\}$ *is a bounded open set. On each one of its connected components* $]a,b[$ *we have*

$$y \leq C_{b-a}^{*\alpha} f(a) = \frac{\alpha}{(b-a)^\alpha}\int_0^{b-a} s^{\alpha-1}\left(f(a+.)\chi_{(0,b-a)}(.)\right)^*(s)ds.$$

Proof : the support of f being bounded and the maximal function taken with respect to $0 < t < T$ it is obvious that $\{M_T f > y\}$ is bounded. The function $C_t^\alpha f(x)$ is continuous so $M_T f(x)$ is lower semi-continuous and $\{M_T f > y\}$ is open.

Let us consider a connected component]a,b[of this open set. For every $u \in]a,b[$ there exists $v > u$ such that

$$C^\alpha_{v-u} f(u) > y.$$

When $\alpha = 1$ it is clear that v can be taken $\leq b$ since $M_T f(b) \leq y$. When $\alpha < 1$ this is not obvious : it is in general false ! This is, at least partly, why we have to consider the $C^{*\alpha}$ instead of C^α in the conclusion. To overcome this difficulty we need the following statement.

Lemma *Let us consider* $a < b < c$, $f \geq 0$ *on* (a,c) *and increasing (wide sense) on* (a,b). *Then we have* $C^\alpha_{c-a} f(a) \leq \sup\left(C^\alpha_{b-a} f(a), C^\alpha_{c-b} f(b)\right)$.

Proof : Let us put

$$\beta(s) = \alpha\left(\frac{b-s}{b-a}\right)^{\alpha-1} \qquad \text{if } a < s < b$$

$$\gamma(s) = \alpha\left(\frac{c-s}{c-a}\right)^{\alpha-1} \qquad \text{if } a < s < c$$

$$\delta(s) = \alpha\left(\frac{c-s}{c-b}\right)^{\alpha-1} \qquad \text{if } b < s < c.$$

For $a < s < b$ the function $r(s) = \gamma(s)/\beta(s) = \left(\frac{b-a}{c-a}\frac{c-s}{b-s}\right)^{\alpha-1}$ is decreasing.

Let us consider the integral $J = \frac{1}{c-a}\int_a^b \gamma(s)\left(\frac{1}{b-a}\int_a^b \beta(u)f(u)du - f(s)\right)ds$

$$= \left(\frac{1}{b-a}\int_a^b \beta(u)\, f(u)du\right)\left(\frac{1}{c-a}\int_a^b \gamma(s)ds\right) - \frac{1}{c-a}\int_a^b \gamma(s)f(s)ds.$$

Let us prove first $J \geq 0$. We write $(c-a)(b-a)J = \int_a^b\int_a^b \gamma(s)\beta(u)(f(u)-f(s))duds$

since $\frac{1}{b-a}\int_a^b \beta(u)du = 1$ which yields

$$(c-a)(b-a)J = \int_a^b \beta(u)\int_u^b \gamma(s)(f(u)-f(s))dsdu + \int_a^b \beta(u)\int_a^u \gamma(s)(f(u)-f(s))dsdu .$$

When $a < u < s < b$, $f(u) - f(s) \leq 0$ and $\gamma(s) = \beta(s)r(s) \leq \beta(s)\, r(u)$. When $a < s < u < b$, $f(u) - f(s) \geq 0$ and $\gamma(s) = \beta(s)\, r(s) \geq \beta(s)\, r(u)$. Therefore we get

$$(c-a)(b-a)\, J \geq \int_a^b \beta(u)\, r(u) \int_u^b \beta(s)\, (f(u) - f(s))\, dsdu + \int_a^b \beta(u)\, r(u) \int_a^u \beta(s)\, (f(u) - f(s))\, dsdu$$

that is $(c-a)(b-a)J \geq -(c-a)(b-a)J$, which proves $J \geq 0$. So we get

$$C^\alpha_{b-a} f(a) = \frac{1}{b-a}\int_a^b \beta(u)\, f(u)du \geq \left(\int_a^b \gamma(u)du\right)^{-1}\int_a^b \gamma(u)f(u)du.$$

Since $\gamma(u)\left(\int_b^c \gamma(s)ds\right)^{-1} = \delta(u)\,(c-b)^{-1}$ when $b < u < c$ we have

$$C^\alpha_{c-b} f(b) = \int_b^c \gamma(u)\, f(u)du\left(\int_b^c \gamma(u)du\right)^{-1} \text{ thus}$$

$$C^{\alpha}_{c-a} f(a) = \frac{1}{c-a}\left[\left(\int_b^c \gamma(u)du\right) C^{\alpha}_{c-b} f(b) + \left(\int_a^b \gamma(u)du\right) C^{\alpha}_{b-a} f(a)\right].$$

Since $\frac{1}{c-a}\int_b^c \gamma(u)du = 1$, this proves the lemma. ♦

Remark : It is easy to see that the assumption "f is increasing on [a,b]" is indispensable.

End of the proof of the rising sun lemma

Let $]a,b[$ be a connected component of the bounded open set $\{M_T f > y\}$. Let $\tilde{f}$ be the function which is equal to f on the complement of $]a,b[$ and to the increasing rearrangement of the restriction of f to $]a,b[$, on $]a,b[$. Since the weight is increasing we have $C^{\alpha}_{v-u}\ \tilde{f}(u) \geq C^{\alpha}_{v-u}\ f(u)$ for every $u \in\ ]a,b[$ and every $v > u$. For every $u \in\ ]a,b[$ there exists $v > u$ such that $C^{\alpha}_{v-u} f(u) > y$ hence $C^{\alpha}_{v-u}\ \tilde{f}(u) > y$. If $v > b$, the lemma implies at once $C^{\alpha}_{b-u}\ \tilde{f}(u) > y$ since, by the definition of $]a,b[$, we have $C^{\alpha}_{v-b}\ \tilde{f}(b) = C^{\alpha}_{v-b}\ f(b) \leq y$. Therefore, for every $u \in\ ]a,b[$ there exists v such that $u < v \leq b$ and $C^{\alpha}_{v-u}\ \tilde{f}(u) > y$. For u fixed, let w be the supremum of such v. By continuity we get $C^{\alpha}_{w-u}\ \tilde{f}(u) \geq y$. If w were such that $w < b$, there should exist $z > w$ with $C^{\alpha}_{z-w}\ f(w) > y$ and $b \geq z$. Since $\tilde{f}$ is increasing on $]a,b[$, as is the weight of the averages we would get, by an application of the corollary 2 of [5] : $C^{\alpha}_{z-u}\ \tilde{f}(u) \geq \frac{w-u}{z-u} C^{\alpha}_{w-u}\ \tilde{f}(u) + \frac{z-w}{z-u} C^{\alpha}_{z-w}\ \tilde{f}(w) > y$,

which would contradict the definition of w. Hence $w = b$, and $C^{\alpha}_{b-u}\ \tilde{f}(u) = C^{*\alpha}_{b-u}\ f(u) \geq y$ for every $u \in\ ]a,b[$. If we let u go down to a, we get the desired result : $C^{*\alpha}_{b-a}\ f(a) \geq y$. ♦

Precise form of the maximal inequality ($0 < \alpha \leq 1$).

For every $f \geq 0$ *measurable on* $\mathbb{R}$ *and* $y > 0$ *we have*

$$\lambda\{Mf > y\} \leq \left(\frac{\alpha}{y}\int_0^{\lambda\{Mf>y\}} t^{\alpha-1}\left(f\chi_{\{Mf > y\}}\right)^*(t)dt\right)^{\frac{1}{\alpha}}.$$

Proof : Let us prove first this inequality for f bounded with compact support and the restricted maximal function M_T. The "sun rising lemma" tells us that $\{M_T f > y\}$ is a denumerable union of pairwise disjoint open intervals $]a,b[$ and on each one we have

$$y \leq \frac{\alpha}{(b-a)^{\alpha}}\int_0^{b-a} t^{\alpha-1}(f\chi_{]a,b[})^*(t)dt.$$

An application of the corollary 2 of **[5]** yields $y \le \frac{\alpha}{\lambda\{Mf>y\}^{\alpha}} \int_0^{\lambda\{Mf>y\}} t^{\alpha-1} (f\chi_{\{Mf>y\}})^*(t)dt$

which is the desired inequality. Taking monotone limits with respect to f and T, we get easily the general case. ♦

Before proceeding to the ergodic setting we are going to discuss briefly the more general case where the averages are taken with respect to an arbitrary increasing weight φ supported in (0,1). It is not difficult to build examples, using only step functions, showing that it is possible to have $f \ge 0$ on (a,c), increasing on (a,b) and also on (b,c) but with averages satisfying

$$A^{*\varphi}(f,(a,c)) \ge A^{\varphi}(f,(a,c)) \ge \sup(A^{*\varphi}(f,(a,b)), A^{*\varphi}(f,(b,c))) \ge \sup(A^{\varphi}(f,(a,b)),A^{\varphi}(f,(b,c))).$$

(the first and the last inequalities being obvious). So the lemma which was the key step in the proof of the "sun rising lemma" does not hold in general. It is the same for the "sun rising lemma" itself. In fact one can show that a necessary condition on φ in order that this lemma holds is the equation $(1-p)\varphi(s) = \varphi\left(\frac{s-p}{1-p}\right)\int_p^1 \varphi(s)ds$ for every $o < p < s < 1$. The only solutions of this equation are $\varphi(s) = \alpha(1-s)^{\alpha-1}$. So the "sun rising lemma" holds only for (C,α)-averages and it is the same for the precise form of the maximal inequality.

Now we come back to the ergodic setting. Given the measurable semi-flow $(\vartheta_s)_{s>0}$ measure preserving on $(\Omega, \mathfrak{F}, \mu)$ the (C,α) averages are $\mathcal{C}_t^{\alpha} f(\omega) = \frac{\alpha}{t^{\alpha}}\int_0^t (t-s)^{\alpha-1} f(\vartheta_s\omega)\, ds$ and the maximal functions are $\mathcal{M} f(\omega) = \sup_{0<t} \mathcal{C}_t^{\alpha} f(\omega)$ or $\mathcal{M}_T f(\omega) = \sup_{0<t<T} \mathcal{C}_t^{\alpha} f(\omega)$.

<u>Precise form of the maximal ergodic inequality</u> $(0 < \alpha \le 1)$

For every $f \ge 0$ *measurable on* Ω *and* $y > 0$ *we have*

$$\mu\{\mathcal{M} f > y\} \le \left(\frac{\alpha}{y}\int_0^{\mu\{\mathcal{M} f>y\}} t^{\alpha-1} (f\chi_{\{\mathcal{M} f>y\}})^*(t)dt\right)^{\frac{1}{\alpha}}$$

<u>Proof</u> : It relies on the transfer principle (**[26] [6]**). For the convenience of the reader we give the details.

We are going to prove the inequality for $\mathcal{M}_T f$. Let us define on the product measure space $(\mathbb{R}_+ \times \Omega, \lambda \otimes \mu)$, the functions $F(x,\omega) = f(\vartheta_x\omega)$ and $F_{u+T}(x,\omega) = f(\vartheta_x\omega)$ if $0 < x < u+T$ and 0 otherwise. We define also $G(x,\omega) = M_T(F(.,\omega))(x)$ and $G_{u+T}(x,\omega) = M_T(F_{u+T}(.,\omega))(x)$ where M_T denote, as before, the maximal operator for the (C,α)-averages, acting on the functions of the real variable x, $F(.,\omega)$ and $F_{u+T}(.,\omega)$ where $\omega \in \Omega$ is supposed to be fixed. We have .

$$C_t^{\alpha}(F(.,\omega))(x) = \mathfrak{C}_t^{\alpha}(f)(\vartheta_x\,\omega)$$

thus $G(x,\omega) = \mathcal{M}_T f(\vartheta_x \omega)$. It is clear that $G_{u+T}(x,\omega) = 0$ for $x > (u+T)$ and $G_{u+T}(x,\omega) = G(x,\omega)$ for $0 < x < u$. Let $E = \{\mathcal{M} f > y\}$ and $\bar{E} = \{(x,\omega)\,;\, G_{u+T}(x,\omega) > y\}$. We have $\bar{E} \subset [0, u+T] \times \Omega$; for a fixed ω, $\bar{E}_\omega = \{x\,;\, G_{u+T}(x,\omega) > y\}$. By Fubini theorem we get

$$\lambda \otimes \mu\,(\bar{E}) = \int_{\mathbb{R}_+} \mu\{\omega\,;\, G_{u+T}(x,\omega) > y\}\, d\lambda(x) \geq \int_0^u \mu\{\omega\,;\, G(x,\omega) > y\}\, d\lambda(x) = u\,\mu(E)$$

because of the invariance of μ under ϑ_x. On the other hand

$$\lambda \otimes \mu\,(\bar{E}) \leq \int_0^{u+T} \mu\{\omega\,;\, G(x,\omega) > y\}\, d\lambda(x) = (u+T)\,\mu(E).$$

Now using the precise form of the maximal inequality for real functions we have

$$y \leq \frac{\alpha}{\lambda\left\{ G_{u+T}(.,\omega)>y \right\}^{\alpha}} \int_0^{\lambda\{Gu+T(.,\omega)>y\}} t^{\alpha-1}\left(F_{u+T}(.,\omega)\, \chi_{\{G_{u+T}(.,\omega)>y\}}\right)^*(t)dt$$

for μ almost every ω in the 2d projection of $\bar{E}$. By the proposition 1 of [5] we get

$$y \leq \frac{\alpha}{(\lambda \otimes \mu(\bar{E}))^{\alpha}} \int_0^{\lambda\otimes\mu(\bar{E})} t^{\alpha-1}(F_{u+T}\chi_{\bar{E}})^*(t)dt\,.$$

Again from the invariance of μ we get $\quad (F_{u+T}\chi_{\bar{E}})^*(t) \leq (f\chi_E)^*(t/_{u+T})$.

Since $u\,\mu(E) \leq \lambda \otimes \mu(\bar{E}) \leq (u+T)\,\mu(E)$ we can write

$$\begin{aligned} y &\leq \frac{\alpha}{\lambda \otimes \mu(\bar{E})} \int_0^{\lambda\otimes\mu(\bar{E})} (f\chi_E)^*(t/_{u+T})\,(t/_{\lambda\otimes\mu(\bar{E})})^{\alpha-1}\, dt \\ &\leq \frac{\alpha}{u\,\mu(E)} \int_0^{(u+T)\mu(E)} (f\chi_E)^*(t/_{u+T})\,\left(t/_{u\,\mu(E)}\right)^{\alpha-1} dt \\ &= \frac{\alpha(u+T)}{u\mu(E)} \int_0^{\mu(E)} (f\chi_E)^*(s)\left(s\,\frac{u+T}{u\mu(E)}\right)^{\alpha-1} ds. \end{aligned}$$

When $u \to +\infty$ we get the desired inequality. The inequality for $\mathcal{M} f$ follows as the limiting case when $T \to \infty$. ♦

A natural question, at this point, is whether the "maximal equality" of Marcus and Petersen [22] holds for (C,α) averages with $0 < \alpha < 1$. An answer would permit an attack of the problem to decide whether the condition $\int_0^1 (f^*(t) \log f^*(t))\, t^{\alpha-1} dt < \infty$ which is sufficient for $\mathcal{M} f \in L(1/\alpha,1)$ (see part 2 above), is also necessary, as it is when $\alpha = 1$, when the flow is ergodic. We have to leave this problem open.

REFERENCES

[1] **N. BINGHAM.** *Moving averages. Almost everywhere convergence I* (ed. G.A. Edgar, L. Sucheston) Academic Press, p. 131-144.

[2] **N. BINGHAM AND L.C.G ROGERS.** *Summability methods and almost sure convergence* (this volume).

[3] **D. BOIVIN AND Y. DERRIENNIC** *The ergodic theorem for cocycles of* $\mathbb{Z}^d$ *or* $\mathbb{R}^d$. (to appear in Ergodic Theory and Dynamical Systems).

[4] **R.P. BOAS.** *A primer of real functions* (The Carus mathematical monographs n° 13) (1960).

[5] **M. BROISE, Y. DÉNIEL AND Y. DERRIENNIC.** *Réarrangement, inégalités maximales et théorèmes ergodiques fractionnaires.* Annales de l'Institut Fourier Tome 39, fasc. 3, p. 689-714 (1989).

[6] **A. CALDERON.** *Ergodic theory and translation invariant operators.* Proc. Nat. Acad. Sci. U.S.A., **59**, p. 349-353 (1968).

[7] **Y.S. CHOW and T.S. LAI.** *Limiting behavior of weighted sums of independent random variables.* Ann. Proba. **1**, p. 810-824 (1973).

[8] **M. de GUZMAN.** *Real variable methods in Fourier analysis.*, North Holland, Math. Studies, **46** (1981).

[9] **Y. DENIEL.** *On the a.e. Cesàro- α convergence for stationary or orthogonal random variables.* Journal of Theoretical Probability, vol **2**, N° **4**, p. 475-485 (1989).

[10] **Y. DENIEL et Y. DERRIENNIC.** *Sur la convergence presque sûre au sens de Cesàro d'ordre* α, $0 < \alpha < 1$ *de v.a. i.i.d.* Proba. theory, **79** p. 629-636 (1988).

[11] **G.H. HARDY.** *Divergent series*, Oxford press, (1959).

[12] **G.H. HARDY et J.E. LITTLEWOOD.** *A maximal theorem with function theoretic applications.* Acta Math. **54**, p. 81-116 (1930).

[13] **G.H. HARDY, J.E. LITTLEWOOD et G. POLYA.** *Inequalities* Cambridge Press, (1934).

[14] **B. HEINKEL.** *An infinite dimensional law of large numbers, in Cesàro's sense.* (to appear).

[15] **R.A. HUNT.** *On* L(p,q) *spaces.* L'enseignement mathématique XII, n° 4 p. 249-276 (1966).

[16] **R. IRMISH.** *Punktweise Ergodensätze für* (C,α)-*Verfahren,* $0 < \alpha < 1$, Dissertation, Fachbereich Math., TH Darmstadt, (1980).

[17] **U. KRENGEL.** *Ergodic theorems*, De Gruyter, (1985).

[18] **T.L. LAI.** *Summability methods for iid random variables*. Proc. Amer. Math. Soc., **43** p. 253-261 (1974).

[19] **G. G. LORENTZ.** *Some new functional spaces*. Ann. of Math., **51**, p. 37-55, (1950).

[20] **G.G. LORENTZ.** *Bernstein polynomials*. Math. Expositions, Toronto press n° 8, 1953.

[21] **G.G. LORENTZ.** *Borel and Banach properties of methods of summation*. Duke Math. J., **n° 22**, p. 129-141, (1955).

[22] **K. PETERSEN.** *Ergodic theory*. Cambridge Univ. Press, (1983).

[23] **E.M. STEIN.** *Singular integrals and differentiability properties of functions*. Princeton Univ. Press, (1970).

[24] **E.M. STEIN and G. WEISS.** *Introduction to Fourier analysis on Euclidean spaces*. Princeton Univ. Press, (1971).

[25] **A.A. TEMPEL'MAN.** *Ergodic theorems for general dynamical systems*. Trudy Moskov. Math. Obsc. 26 = Trans. Moscow Math. Soc. **26,** p. 94-132, (1972).

[26] **N. WIENER.** *The ergodic theorem*, Duke Math. J., **5**, p. 1-18, (1939).

[27] **A. ZYGMUND.** *Trigonometric series,* Cambridge Univ. Press, (1959).

M. BROISE, Y. DÉNIEL, Y. DERRIENNIC
Département de Mathématiques et Informatique
Université de Bretagne Occidentale
6, Avenue Victor Le Gorgeu
29287 Brest (France).

* Vladimir Gaposhkin informed the authors that the (C, 1/2)-strong law of large numbers for i.i.d. square integrable variables can also be deduced from one of his results.

THE HILBERT TRANSFORM OF THE GAUSSIAN

A.P. Calderón and Y. Sagher

Dedicated to the memory of N.M. Rivière.

In this short note we calculate the Hilbert transform of the Gaussian. The problem was first mentioned to one of us some twenty years ago, by N. M. Rivière. As it turns out, the solution is very natural, leading to a well known function. We will give two proofs of the result, the second much shorter than the first. On the other hand the first proof yields some additional results. It also suggests a connection between the Hilbert transform and the one dimensional heat equation. For some applications of this connection see [1], where the Hilbert transforms of the Hermite functions $H_n(x)\mathcal{G}(x)$ are also calculated.

We begin by considering a classical analog of the problem, Laplace's equation for the upper half plane. Consider kernels $\phi(x)$ so that $\phi_y(x) = \frac{1}{y}\phi(\frac{x}{y})$ satisfies $\Delta\phi = \partial_x^2\phi + \partial_y^2\phi = 0$. This leads to the differential equation:

$$(1 + u^2)\phi''(u) + 4u\phi'(u) + 2\phi(u) = 0 \qquad (1)$$

which can be rewritten:

$$\frac{d}{du}\left[\frac{d}{du}\left((1 + u^2)\phi(u)\right)\right] = 0 \qquad (2)$$

The general solution of this equation is $\phi(u) = \dfrac{A}{1 + u^2} + \dfrac{Bu}{1 + u^2}$. Taking $A = \frac{1}{\pi}$ and $B=0$ gives P(u), Poisson's kernel for the upper half plane. Taking $A=0$ and $B = \frac{1}{\pi}$ gives Q(u), the conjugate Poisson kernel for the upper half plane. If we denote by H the Hilbert transform:

$$H(f)(x) = \text{p.v. } \frac{1}{\pi}\int_{\mathbb{R}} \frac{f(x-t)}{t}dt \qquad (3)$$

then, as is well known, $H(P) = Q$.

ISBN 0-12-085520-8

The argument above serves as motivation for what we do next.

Let T be the heat differential operator:

$$T = \partial_x^2 - \partial_t \tag{4}$$

and consider functions $\phi(u)$ which satisfy $T(\phi_{\sqrt{2t}}(x)) = 0$. This leads to the differential equation

$$\phi''(u) + u\phi'(u) + \phi(u) = 0 \tag{5}$$

The equation can be easily transformed to:

$$\frac{d}{du}\left[e^{-u^2/2}\cdot\frac{d}{du}\left(e^{u^2/2}\phi(u)\right)\right] = 0 \tag{6}$$

The general solution of the last equation is:

$$\phi(u) = Ae^{-u^2/2} + Be^{-u^2/2}\int_0^u e^{s^2/2}ds \tag{7}$$

Taking $A=\dfrac{1}{\sqrt{2\pi}}$, B=0 gives us the Gaussian, $\mathcal{G}(u)$. We denote:

$$\mathcal{S}(u) = \frac{1}{\pi}e^{-u^2/2}\int_0^u e^{s^2/2}ds \tag{8}$$

and prove $H(\mathcal{G}) = \mathcal{S}$.

A simple application of l'Hôpital's proves:

LEMMA 1. As $|u| \longrightarrow \infty$,

(i) $e^{-u^2/2}\int_0^u e^{s^2/2}(u-s)^2ds = O((\frac{1}{u})^3)$

(ii) $\mathcal{S}(u) = O(\frac{1}{u})$.

THEOREM 1. For $f\in L^p$, $1 \le p < \infty$, $\lim_{\varepsilon\to 0} \mathcal{S}_\varepsilon * f = H(f)$ almost everywhere. The convergence is in L^p norm if $1<p<\infty$, and in $L(1,\infty)$ quasi-norm if $f\in L^1$.

PROOF: $\pi\mathcal{S}(u) = e^{-u^2/2}\int_0^u e^{s^2/2}ds = e^{-u^2/2}\int_0^u e^{(u-s)^2/2}ds = \int_0^u e^{-us}e^{s^2/2}ds =$

$$= \int_0^u e^{-us} \cdot \sum_{j=0}^{\infty} \frac{1}{j!}\left(\frac{s^2}{2}\right)^j ds = \frac{1}{u}(1 - e^{-u^2}) + \int_0^u e^{-us} \cdot \sum_{j=1}^{\infty} \frac{1}{j!}\left(\frac{s^2}{2}\right)^j ds.$$

Denote: $k(u) = \left\{ \begin{array}{ll} \frac{1}{\pi u} & \text{if } |u|>1 \\ 0 & \text{otherwise} \end{array} \right\}$. Let $\Psi(u) = \mathcal{S}(u) - k(u)$. For $u > 1$,

$$\pi|\Psi(u)| \le e^{-u^2} + \int_0^u e^{-us}\left(\frac{s^2}{2}\right)\sum_{j=0}^{\infty} \frac{1}{j!}\left(\frac{s^2}{2}\right)^j ds \le$$

$$\le e^{-u^2} + \int_0^u e^{-us}\left(\frac{s^2}{2}\right)e^{s^2/2} ds = e^{-u^2} + e^{-u^2/2}\int_0^u e^{(u-s)^2/2}\left(\frac{s^2}{2}\right)ds =$$

$$= e^{-u^2} + \frac{1}{2}e^{-u^2/2}\int_0^u e^{s^2/2}(u-s)^2 ds \le \frac{c}{u^3}.$$

Since ψ is an odd function, $|\psi(u)| \le \dfrac{c}{|u|^3}$ for all $|u| \ge 1$. For $|u| \le 1$, $\psi = \mathcal{S}$ and so $\pi|\psi(u)| \le 1$. In sum: $|\psi|$ is bounded by a bell-shaped L^1 function. This proves that for all functions $f \in L^p$, $1 \le p < \infty$, $\|\psi_\varepsilon * f\|_p \longrightarrow 0$ as $\varepsilon \longrightarrow 0$, as well as $\psi_\varepsilon * f \longrightarrow 0$ at all Lebesgue points of f. Since $\lim_{\varepsilon \to 0} h_\varepsilon * f = H(f)$, at all Lebesgue points, in L^p norm for $1 < p < \infty$ and in $L(1,\infty)$ quasi-norm for p=1, we have $\lim_{\varepsilon \to 0} \mathcal{S}_\varepsilon * f = H(f)$ at all Lebesgue points, in L^p norm if $f \in L^p$ and $1 < p < \infty$, and in $L(1,\infty)$ quasi-norm if $f \in L^1$. ▮

THEOREM 2. $H(\mathcal{G}) = \mathcal{S}$.

PROOF. From Lemma 1 follows that $\mathcal{S} \in L^p$ for all $1<p$. The convolutions appearing in the proof below are therefore all well defined.

For any $f \in L^p$, $1<p<\infty$, we have $\mathcal{S}_\varepsilon * H(f) \longrightarrow H(H(f)) = -f$. On the other hand $\mathcal{S}_\varepsilon * H(f) = H(\mathcal{S}_\varepsilon) * f$ and $H(\mathcal{S}_\varepsilon) = (H(\mathcal{S}))_\varepsilon$ so that we have: $(H(\mathcal{S}))_\varepsilon * f = H(\mathcal{S}_\varepsilon) * f \longrightarrow -f$. Since $T(\mathcal{S}_{\sqrt{2t}}) = 0$, it can easily be proved that $T(H(\mathcal{S}_{\sqrt{2t}})) = 0$. $H(\mathcal{S})$ therefore satisfies equation (5) above and so $H(\mathcal{S}) = A\mathcal{G} + B\mathcal{S}$. $\mathcal{S}$ however is an odd function so that $H(\mathcal{S})$ is an even function, as is $\mathcal{G}$. Therefore B=0. Since both $H(\mathcal{S})_\varepsilon * f \longrightarrow -f$ and $\mathcal{G}_\varepsilon * f \longrightarrow f$, we have A= -1, and $H(\mathcal{S}) = -\mathcal{G}$. Applying H to both sides of the equation gives $H(\mathcal{G}) = \mathcal{S}$. ▮

For some results concerning the function $\mathscr{S}$ see [3].

Let us present a second proof of $H(\mathscr{G}) = \mathscr{S}$.

Denote $f^{\wedge}(x) = \frac{1}{\sqrt{2\pi}}\int_{\mathbb{R}} f(t)e^{-ixt}dt$. Then $\mathscr{G}^{\wedge}(x) = \mathscr{G}(x)$, and $(Hf)^{\wedge}(x) = -i\,\mathrm{sign}\,x\cdot f^{\wedge}(x)$.

Therefore:

$$H(\mathscr{G})(t) = -\frac{i}{\sqrt{2\pi}}\int_{\mathbb{R}} \mathrm{sign}\,x\cdot\mathscr{G}(x)e^{ixt}dx = \frac{2}{\sqrt{2\pi}}\mathrm{Im}\left(\int_0^\infty \mathscr{G}(x)e^{ixt}dx\right) = \frac{1}{\pi}e^{-t^2/2}\mathrm{Im}\left(\int_0^\infty e^{-(it-x)^2/2}dx\right).$$

Consider x now as a complex variable, x=u+iv. Clearly, $\int_\Gamma e^{-(it-x)^2/2}dx = 0$, where Γ is the contour:

(contour: from 0 up to it along Γ_2, then along Γ_1 parallel to the real axis)

Since $\int_{\Gamma_1} e^{-(it-x)^2/2}dx$ is real, we have:

$$\frac{1}{\pi}e^{-t^2/2}\mathrm{Im}\left(\int_0^\infty e^{-(it-x)^2/2}dx\right) = \frac{1}{\pi}e^{-t^2/2}\mathrm{Im}\left(i\int_0^t e^{-(it-iv)^2/2}dv\right) = \frac{1}{\pi}e^{-t^2/2}\int_0^t e^{x^2/2}dx. \blacksquare$$

Still another proof of Theorem 2 will appear in [2].

BIBLIOGRAPHY

1. E. Kochneff, Y. Sagher, K. Zhou, Homogeneous solutions of the heat equation. To appear.
2. E. Kochneff, Y. Sagher, R. Tan, On a theorem of Akhiezer, to appear.
3. Lebedev, N.N., Special Functions and Their Applications. Prentice Hall, 1965.

University of Chicago.

University of Illinois, Chicago.
Syracuse University.

Mean Ergodicity of L_1 Contractions and Pointwise Ergodic Theorems

Doḡan Çömez
North Dakota State University, Fargo ND 58105, U.S.A.

Michael Lin
Ben Gurion University of the Negev, Beer Sheva, Israel

Abstract

Let T be a (linear) contraction on L_1 with modulus $\mathcal{T}$. We show that if $\mathcal{T}$ is mean ergodic, so is T, and the averages $A_n(T)f$ and $A_n(\mathcal{T})f$ converge a.e. for any $f \in L_1$. The mean ergodicity of $\mathcal{T}$ is not necessary for the mean ergodicity of T together with the a.e. convergence of the averages. If $\{F_n\}$ if a dominated super additive process with respect to a mean ergodic positive contraction T, then $\frac{1}{n}F_n$ converges a.e. and in L_1, to a T-invariant function.

Let $T_1, ..., T_d$ be commuting contractions, and let $A_n(T_1, ..., T_d) = A_n(T_1)...A_n(T_d)$. Then $A_n(T_1, ..., T_d)f$ converges in L_1 for every $f \in L_1$ if the associated Brunel operator is mean ergodic. If the T_i are *positive*, the converse is also true. We prove that if the moduli are mean ergodic and commute, the averages $A_n(T_1, \ldots, T_d)f$ and $A_n(\mathcal{T}_1, \ldots, \mathcal{T}_d)f$ converge a.e. for every $f \in L_1$.

0 Introduction

Let T be a contraction of both L_1 and L_∞ of a σ-finite measure space (X, Σ, μ). The celebrated theorem of Dunford and Schwartz says that the averages $A_n(T)f = (\frac{1}{n}\sum_{j=0}^{n-1} T^j)f$ converge a.e., for any $f \in L_1$.

If T is a *positive* contraction of $L_1(X, \Sigma, \mu)$ such that $Tu \leq u$ a.e. for some (not necessarily integrable) function u with $0 < u < \infty$ a.e., then (by changing to the measure $d\nu = ud\mu$ and applying the above result) $A_n(T)f$ converges a.e. for any $f \in L_1$.

ISBN 0-12-085520-8

For T *positive* and μ *finite*, Ito [I_1] generalized the Dunford-Schwartz theorem (Hopf's in that case) by showing that a.e. convergence holds if $\{A_n(T)1\}$ is uniformly integrable. If, for some $w \in L_1$ with $w > 0$ a.e., $\{A_n(T)w\}$ is uniformly integrable, then (by looking at $d\nu = wd\mu$ and using Ito's result), $A_n(T)f$ converges a.e. for any $f \in L_1$ (Kim [Ki]). This last assumption is in fact equivalent to *mean ergodicity* of the positive contraction T:

(i) If, for $f \in L_1$ $\{A_n(T)f\}$ is uniformly integrable, it is weakly sequentially compact, hence [K, p.72] converges in norm.

(ii) If $\{A_n(T)w\}$ is uniformly integrable, for $w \in L_1$ with $w > 0$ a.e., then so is $A_n(T)f$ for any $f \in L_1$ with $|f| \leq cw$. Continuity of T now yields the mean ergodicity.

For other equivalent conditions, see [K, p.174-175]. In general, mean ergodicity does not imply the Dunford-Schwartz condition.

Clearly on the dissipative part D, $A_n(T)f \to 0$ a.e. on D for any $f \in L_1$. Weiner's example [W] shows that T may be conservative (and ergodic) with $A_n(T)f$ converging a.e., without being mean ergodic, nor satisfying the Dunford-Schwartz condition (there is no $u \geq 0$ finite a.e. with $Tu \leq u$, except $u \equiv 0$). See also [I_2].

In section 1 we extend Kim's result to non-positive contractions, and obtain a.e. and L_1-convergence for dominated superadditive processes with respect to a positive mean ergodic contraction. In section 2 we discuss the mean ergodicity of a discrete d-parameter semi-group generated by d commuting contractions. We obtain a multi-dimensional extension of Kim's result, even for non-positive contractions. This generalizes the multi-dimensional Dunford-Schwartz theorem (see also [B]) for *finite* μ. (see [M] for another direction of generalization).

1 On Ergodic Convergence for a General L_1 Contraction

In this section we study the convergence properties of the averages $A_n(T) = \frac{1}{n}\sum_{k=0}^{n-1} T^k$ of a (linear) contraction T on L_1.

We start with an example, showing that, unlike the positive case proved by Ito-Kim, mean ergodicity of T does not imply a.e. convergence of $A_n(T)f$, $f \in L_1$.

Example Let $\mathcal{T}$ be the positive contraction of L_1 constructed by Chacon (see [K,p.151]). There is $f \in L_1^+$ for which $\limsup \frac{1}{n}\mathcal{T}^{n-1}f = \infty$ a.e. Let $T = -\mathcal{T}$, and $g = h + \mathcal{T}h$. Then

$$A_{2n}(T)g = \frac{1}{2n}\sum_{k=0}^{2n-1}(-\mathcal{T})^k g = \frac{1}{2n}\sum_{k=0}^{n-1}(\mathcal{T}^{2k} - \mathcal{T}^{2k+1})g =$$

$$\frac{1}{2n}\sum_{k=0}^{n-1}\mathcal{T}^{2k}(h - \mathcal{T}^2 h) = \frac{1}{2n}(h - \mathcal{T}^{2n}h).$$

Since $\{\mathcal{T}^{n-1}f/n\}$ is unbounded a.e., either there exists a set of positive measures on which $\{\mathcal{T}^{2n}f/2n\}$ is unbounded, or a set on which $\{\mathcal{T}^{2n+1}f/(2n+1)\}$ is unbounded. In the first case we take $h = f$, in the second $h = \mathcal{T}f$, to obtain that $A_{2n}(T)g$ does not converge on a set of positive measure. Since in the construction not only $\mathcal{T}$ is ergodic, but also $\mathcal{T}^2$, $T^*f = f \in L_\infty \Leftrightarrow \mathcal{T}^*f = -f$ holds only for $f = 0$. Hence $\|A_n(T)f\|_1 \to 0$ for every $f \in L_1$ by the mean ergodic theorem.

Our study will involve the linear modulus $\mathcal{T}$ of a contraction on L_1 [K, p.159], and its conservative part C and dissipative part D.

Proposition 1.1 *Let T be a contraction on L_1, with modulus $\mathcal{T}$. If $\mathcal{T}$ is mean ergodic, then T is mean ergodic, and for every $f \in L_1$, $A_n(T)f$ and $A_n(\mathcal{T})f$ converge a.e.*

Proof: By Helmberg's criterion [K,p.175], $\mathcal{T}$ has a fixed point $g \in L_1^+$, with $\{g > 0\} = C$, and $\mathcal{T}^{*n}1_D \downarrow 0$ a.e. By the Chacon-Ornstein theorem,

$A_n(\mathcal{T})f$ converges a.e. on C for any $f \in L_1$. By the definition of D, $A_n(\mathcal{T})f \to 0$ a.e. on D.

Since $|A_n(T)f| \leq A_n(\mathcal{T})|f|$, uniform integrability of $A_n(T)f$ follows, and (by the Yosida-Kakutani mean ergodic theorem) $A_n(T)f$ converges in L_1 norm.

By Chacon's general ratio theorem ([K, p.164], with $p_n = g$), $A_n(T)f$ converges a.e. on C for $f \in L_1$, and clearly $|A_n(T)f| \leq A_n(\mathcal{T})|f|$ converges to 0 a.e. on D.

Remark $\mathcal{T}$ in the previous example cannot be mean ergodic, by proposition 1.1. Hence $T^2 = \mathcal{T}^2$ is not mean ergodic, although T is mean ergodic. (Note that if T is a mean ergodic *positive* contraction, T^m is also mean ergodic).

Lemma 1.2 *Let T be a contraction of L_1 with modulus $\mathcal{T}$. If $Th = h \in L_1$, then $\{h \neq 0\} \subset C$, and $\mathcal{T}|h| = |h|$.*

Proof: We have $|h| = |Th| \leq \mathcal{T}|h|$, and since $\mathcal{T}$ is a contraction, $\mathcal{T}|h| = |h|$. Hence $\{h \neq 0\} = \{|h| > 0\} \subset C$.

Theorem 1.3 *Let T be a contraction in L_1 with modulus $\mathcal{T}$. If there exists $h \in L_1$ with $Th = h$, and $|h| > 0$ a.e., then:*

(i) *Both T and $\mathcal{T}$ are mean ergodic.*

(ii) *For every $f \in L_1$, $A_n(T)f$ and $A_n(\mathcal{T})f$ converge a.e.*

Proof: By lemma 1.2, $\mathcal{T}$ has a strictly positive fixed point, so $D = \emptyset$, and $\mathcal{T}$ is mean ergodic [K, p.128]. Proposition 1.1 yields the other assertions.

The following example shows that the converse of proposition 1.1 is false.

Example T a mean ergodic contraction on L_1 such that $A_n(T)f$ and $A_n(\mathcal{T})f$ converge a.e. for every $f \in L_1$, but $\mathcal{T}$ is *not* mean ergodic.

Let $\mathcal{T}'$ be a conservative and ergodic contraction on L_1 with a σ-finite *infinite* invariant measure, and let $\mathcal{T} = \frac{1}{2}(I + \mathcal{T}')$. Let $T = -\mathcal{T}$. Then

$\mathcal{T} = |T|$. $A_n(\mathcal{T})f$ converges to zero a.e. for every $f \in L_1$, and $|A_n(T)f| \le A_n(\mathcal{T})|f| \to 0$ a.e. Now $I - T = I + \mathcal{T} = \frac{1}{2}(3I + \mathcal{T}')$ is invertible, so $\|A_n(T)\| \to 0$. Finally, since $\mathcal{T}$ has no finite invariant measures, $\mathcal{T}$ is not mean ergodic ($A_n(\mathcal{T})f$ can converge in L_1 norm only to zero, which happens if and only if $\int f = 0$).

Theorem 1.4 *Let T be a contraction in L_1 (with modulus $\mathcal{T}$). If there exists $h \in L_1$ with $Th = h$, and $|h| > 0$ a.e. on C, then for every $f \in L_1$, the averages $A_n(T)f$ (and $A_n(\mathcal{T})f$) converge a.e.*

Proof: By lemma 1.2, $\mathcal{T}|h| = |h|$, and $\{|h| > 0\} = C$. The convergence follows from the proof of proposition 1.1

Remarks 1. Theorem 1.4 provides an extension of Kim's theorem to the non-positive case, expressed in direct terms of T (while the extension in proposition 1.1 uses $\mathcal{T}$.)
2. A.e. convergence in proposition 1.1 is not true if T is assumed power bounded instead of a contraction. For a counter-example (with T *positive*) of $f \in L_1$ for which $A_n(T)f$ does not converge a.e. see [DL].
Definition A sequence $\{F_n\} \subset L_1$ is a *superadditive* process with respect to a positive L_1 contraction if $F_0 = 0$ and $F_{n+m} \ge F_n + T^n F_m$ for every $n, m \ge 0$.

A superadditive process F_n is *bounded* if $\sup\{\frac{1}{n}\|F_n\|_1 : n \ge 1\}$ is finite. It is *dominated* if there exists $g \in L_1$ (called a *dominant*) such that $F_n \le \sum_{i=0}^{n-1} T^i g$ for every $n \ge 1$. The dominant g is *exact* if $\lim_{n\to\infty} \frac{1}{n}\|F_n\|_1 = \lim \|\frac{1}{n}\sum_{i=0}^{n-1} T^i g\|_1$ (both limits exist - see [K, p.36])

Theorem 1.5 *Let T be a mean ergodic positive L_1 contraction, and let $\{F_n\}$ be a dominated super additive process. Then $\frac{1}{n}F_n$ converges a.e. and in L_1, and the limit is T-invariant.*

Proof: T has an equivalent finite invariant measure on its conservative part C. By looking at $F_n' = F_n - \sum_{i=0}^{n-1} T^i F_1$, we may and do assume that

$\{F_n\}$ is positive, since $\frac{1}{n}\sum_{i=0}^{n-1} T^i F_1$ converges in L_1 and a.e. by proposition 1.1.

Now $L_1(C)$ is T-invariant, and $H_n = I_C F_n$ is a positive bounded superadditive process. Also, T_C is conservative, and has a strictly positive fixed point, so a.e. convergence of $I_C F_n/n$ follows from the ratio ergodic theorem for superadditive processes of Akcoglu and Sucheston [AS] (see also [K, p.149]). Hence F_n/n converges a.e. on C.

The domination of $\{F_n\}$ yields that $0 \le F_n/n \le A_n(T)g \to 0$ a.e. on D, (where g is a dominant of the positive $\{F_n\}$).

We now prove L_1-convergence, and invariance of the limit. We assume first $X = C$ (i.e., we look at $H_n = I_C F_n$ as before). In that case, F_n has an *exact dominant* [AS] g, so $A_n(T)g - \frac{1}{n}F_n \ge 0$, and, since g is exact and T Markovian on C,

$$\|A_n(T)g - \frac{1}{n}F_n\|_1 = \int g d\mu - \frac{1}{n}\|F_n\|_1 \to 0$$

Hence on C we have L_1-convergence to $\bar{g} = \lim A_n(T)g$ (which exists by the ergodic theorem), and $T\bar{g} = \bar{g}$.

Now, since $\bar{g} = 0$ on D, $\frac{1}{n}F_n \to 0$ a.e. on D and $F_n \ge 0$,

$$\|\bar{g} - \frac{1}{n}F_n\|_1 = \int_C |\bar{g} - \frac{1}{n}F_n| d\mu + \int_D \frac{1}{n}F_n d\mu = \|\bar{g} - \frac{1}{n}H_n\|_1 + \int_D \frac{F_n}{n} d\mu \to 0,$$

using the domination for the second limit.

Remarks 1. If T is *Markovian* (i.e., $T^*1 = 1$), and $\{F_n\}$ is bounded, then it is dominated, and has an exact dominant on the whole space [AS]. However, we do not require T to be Markovian.

2. Brunel and Sucheston [BS] proved that $\{F_n\}$ has an exact dominant if and only if

$$\lim_{N\to\infty} \frac{1}{N}\sum_{n=1}^{N} \|F_n - TF_{n-1} - F_1\| < \infty,$$

i.e., if and only if $\liminf \frac{1}{N}\sum_{n=1}^{N} \|F_n - TF_{n-1}\| < \infty$.

3. The T-invariance of the limit follows also from [AS] (the case of a finite invariant measure and L_1-convergence is not considered there).

For the sake of completeness we include the following.

Theorem 1.6 *Let T be a positive $L_1 - L_\infty$ contraction, and let $\{F_n\}$ be a dominated superadditive process. Then $\frac{1}{n}F_n$ converges a.e.*

Proof: Since $A_n(T)g$ converges a.e. for any $g \in L_1$, we may assume $F_n \geq 0$. A.e. convergence on D follows from the domination, as before. $I_C F_n$ is now a bounded positive superadditive process with respect to T_c, so we may assume $X = C$. Then there is an exact dominant h, and

$$\frac{1}{n}F_n = \frac{F_n}{\sum_{k=0}^{n-1} T^k h} A_n(T)h \text{ on } \{\sum_{k=0}^{\infty} T^k h > 0\}$$

yields the a.e. convergence on that set, by [AS]. On the complement $F_n = 0$, so convergence holds a.e.

Remark The above idea is used in [BS] for obtaining a general ratio ergodic theorem. Theorem 1.6 can be also deduced from that result.

2 Ergodic Convergence of Commuting L_1 Contractions

We denote $A_n(T_1, ..., T_d) = A_n(T_1)A_n(T_2)...A_n(T_d)$.

Proposition 2.1 *Let $T_1, ..., T_d$ be commuting positive contractions on L_1. If for some $w \in L_1^+$ with $w > 0$ a.e. the sequence $A_n(T_1, ..., T_d)w$ is uniformly integrable, then for every $f \in L_1$, the sequence $A_n(T_1, ..., T_d)f$ converges in L_1 norm, and the limit f^* satisfies $T_i f^* = f^*$ for every $1 \leq i \leq d$.*

Proof: Since $(I - T_j)A_{n_k}(T_j)f \to_{k\to\infty} 0$, and $\{T_i\}$ commute, the weak sequential compactness (equivalent to uniform integrability) of $A_n(T_1, ..., T_d)f$ implies, by Eberlein's ergodic theorem [K, p.76], that $A_n(T_1, ..., T_d)f$ converges in norm, to a common invariant function f^*.

We thus have convergence of $A_n(T_1, ..., T_d)f$ for $f \in L_1$ with $|f| \leq cw$, since

$$|A_n(T_1, ..., T_d)f| \leq A_n(T_1, ..., T_d)|f| \leq cA_n(T_1, ..., T_d)w$$

yields uniform integrability of the sequence. By continuity, the propositon is proved.

Remark The converse implication is trivial.

A useful tool for investigating finitely many commuting contractions is the Brunel operator U introduced in [B] (see [K, p.213]).

Theorem 2.2 *Let $T_1, ..., T_d$ be commuting contractions of L_1, and let U be the corresponding Brunel operator.*

(i) *If U is mean ergodic, $A_n(T_1^m, ..., T_d^m)f$ converges in L_1-norm, for m fixed and every $f \in L_1$.*

(ii) *If the moduli $\mathcal{T}_i$ commute, then U is mean ergodic if and only if $A_n(\mathcal{T}_1, ..., \mathcal{T}_d)f$ converges in L_1-norm, for every $f \in L_1$.*

Proof: (i) Brunel's estimate [K, p.213]

$$|A_n(T_1, ..., T_d)f| \leq cA_{n_d}(U)|f|$$

yields the uniform integrability of $\{A_n(T_1, ..., T_d)f\}$, since $\{A_{n_d}(U)|f|\}$ is norm convergent. The desired convergence follows, as in proposition 2.1, by Eberlein's theorem. We now prove the case $m > 1$. Since

$$|\sum_{j=0}^{n-1}(T_i^m)^j f| \leq \sum_{j=0}^{n-1}(\mathcal{T}_i^m)^j|f| \leq \sum_{k=0}^{mn-1}\mathcal{T}_i^j|f|$$

we obtain that for an appropriate $k = k(n)$, by Brunel's estimate,

$$|\frac{1}{m^d}A_n(T_1^m, \ldots, T_d^m)f| \leq A_{mn}(\mathcal{T}_1)\ldots A_{mn}(\mathcal{T}_d)|f| \leq CA_k(U)|f|,$$

(see [K, p.214] for the construction which yields the inequality even if the moduli $\{\mathcal{T}_i\}$ do not commute).

Since U is mean ergodic, the sequence $\{A_n(T_1^m, \ldots, T_d^m)f\}_{n=1}^{\infty}$ is uniformly integrable, and Eberlein's theorem concludes the proof as before.

(ii) If U is mean ergodic, the desired convergence follows from (i), as U constructed for $\{T_i\}_{i=1}^d$ is the same as for $\{\mathcal{T}_i\}_{i=1}^d$, since they commute.

We now assume that $A_n(\mathcal{T}_1, ..., \mathcal{T}_d)f$ converges for every $f \in L_1$. By the "splitting theorem" [K, p.77],

$$L_1 = \{h \in L_1 : \mathcal{T}_i h = h \text{ for } 1 \leq i \leq d\} \oplus clm\{\cup_{i=1}^d (I - \mathcal{T}_i)L_1\}$$

Since $\{\mathcal{T}_i\}_{i=1}^d$ commute, $Uh = h \in L_1$ if and only if $\mathcal{T}_i h = h$ for every i, and $U^*\phi = \phi \in L_\infty$ if and only if $\mathcal{T}_i^*\phi = \phi$ for every i, by the Brunel-Falkowitz lemma [K, p.82]. Let $U^*\phi = \phi \neq 0$. Then $\mathcal{T}_i^*\phi = \phi$, so $\phi = 0$ on $clm\{\cup_{i=1}^d (I - \mathcal{T}_i)L_1\}$. Hence there exists $h \neq 0$ with $Uh = h$ and $\langle h, \phi \rangle \neq 0$.

Thus, the fixed points of U separate those of U^*, and by Sine's criterion [K, p.74], U is mean ergodic.

Remarks 1. By proposition 2.1, we may use in (ii) also the condition: "*for some $w \in L_1$ with $w > 0$ a.e. $\{A_n(\mathcal{T}_1, ..., \mathcal{T}_d)w\}$ is uniformly integrable.*"

2. In part (i) the moduli $\mathcal{T}_i$ need *not* commute.

3. For $d = 1$, we see that mean ergodicity of $\mathcal{T}$ is equivalent to that of U (a convex combination of $\{\mathcal{T}^n : n \geq 0\}$ with strictly positive coefficients). Thus, theorem 2.2 suggests that extensions of the Ito-Kim result could be obtained by conditions on U.

Lemma 2.3 *Let $T_1, \ldots, T_d$ be commuting mean ergodic contractions on a Banach space X, with $E_j = \lim_{n\to\infty} A_n(T_j)$. Then, for every $x \in X$,* $\lim_{n\to\infty} \|A_n(T_1, \ldots, T_d)x - E_1E_2 \ldots E_d x\| = 0$.

Proof: Since the T_j commute, so do the projections E_j.

The lemma is proved by induction on the number of operators:

$$\|A_n(T_1, \ldots, T_d)x - E_1 \ldots E_d x\| \leq$$

$$\|A_n(T_1, \ldots, T_d)x - A_n(T_1, \ldots, T_{d-1})E_d x\| +$$
$$\|(A_n(T_1, \ldots, T_{d-1}) - E_1 \ldots E_{d-1})E_d x\|$$

$$= \|A_n(T_1, \ldots, T_{d-1})(A_n(T_d)x - E_d x)\| +$$

$$\|(A_n(T_1, \ldots, T_{d-1}) - E_1 \ldots E_{d-1})E_d x\| \to 0,$$

using mean ergodicity of T_d and the induction hypothesis.

Theorem 2.4 *Let $T_1, ..., T_d$ be commuting contractions of L_1 and let U be the associated Brunel operator. If U is also an L_∞-contraction, then for every $f \in L_1$, the sequence $A_n(T_1, ..., T_d)f$ converges a.e.*

Proof: Brunel's proof (see [K, p.215]) uses only the estimates by $A_n(U)$ and the fact that U is an $L_1 - L_\infty$ contraction.

Theorem 2.5 *Let $T_1, ..., T_d$ be commuting contractions of L_1. If there exists $w \in L_1$, $w > 0$ a.e. and $Uw = w$, then for every $f \in L_1$ the sequence $A_n(T_1, ..., T_d)f$ converges a.e. and in L_1.*

Proof: U is mean ergodic in L_1, having an equivalent finite invariant measure, so L_1-convergence follows from theorem 2.2. By changing the measure to $d\mu' = w d\mu$, we obtain a.e. convergence from theorem 2.4.

Corollary 2.6 *Let $T_1, ..., T_d$ be commuting contractions of L_1. If there exists $h \in L_1$ with $|h| > 0$ a.e., and $T_i h = h$ for $1 \le i \le d$, then for every $f \in L_1$, the sequence $A_n(T_1, ..., T_d)f$ converges a.e. and in L_1.*

Proof: By lemma 1.2, $\mathcal{T}_i|h| = |h|$ for each i, and by the definition of U, $U|h| = |h|$, so theorem 2.5 applies.

Remark This corollary follows also directly from Brunel's theorem, by a change of measure.

Our main result generalizes Proposition 1.1 to d dimensions. We will need mean ergodicity and *commutativity* of the moduli $\mathcal{T}_i$.

Lemma 2.7 *Let $T_1, \dots, T_d$ be commuting mean ergodic contractions on a Banach space X. Then there exist operators $R_i = \sum_j B_{ij}$, where each B_{ij} is of the form $B_{ij} = \pm\Pi_{k=1}^d S_k$ with $S_k \in \{I, T_k\}$, such that the set*

$$Y = \{\sum_{i=1}^{d} R_i y_i + \Pi_{k=1}^d (I - T_k)z : z \in X,\ T_i y_i = y_i \text{ for } 1 \le i \le d\}$$

is dense in X.

Proof: By induction on d. For $d = 1$, $X = \{y : Ty = y\} \oplus \overline{(I-T)X}$ by mean ergodicity of T.

In order to illustrate the passage from $d-1$ to d operators, we first show the case $d = 2$. By mean ergodicity of T_2, $\{y_2 : T_2 y_2 = y_2\} + (I - T_2)X$ is dense.

Since $\{y_1 : T_1 y_1 = y_1\} + (I - T_1)X$ is dense, we have the density of the set $\{y_2 + (I - T_2)[y_1 + (I - T_1)z] : T_i y_i = y_i, z \in X\}$.

But then $R_2 = I$, $R_1 = I - T_2$ yield the assertion.

Assume the lemma is true for $d - 1$ operators. Since T_d is mean ergodic the set $\{y_d : T_d y_d = y_d\} + (I - T_d)X$ is dense.

By the induction hypothesis, we obtain the density of the set

$$\{y_d + (I - T_d)[\sum_{i=1}^{d-1} R_i^{(d-1)} y_i + \Pi_{k=1}^{d-1}(I - T_k)z] : z \in X,\ T_i y_i = y_i \text{ for } 1 \le i \le d\}$$

Hence we take $R_d^{(d)} = I$, and $R_i^{(d)} = (I - T_d)R_i^{(d-1)}$, so that $B_{ij}^{(d)}$ is $B_{ij}^{(d-1)}$ or $-T_d B_{ij}^{(d-1)}$, which is of the required form.

Theorem 2.8 *Let $T_1, \dots, T_d$ be commuting L_1-contractions whose moduli $\mathcal{T}_1, \dots, \mathcal{T}_d$ are mean ergodic and commute. Then for every $f \in L_1$, the sequences $A_n(T_1, \dots, T_d)f$ and $A_n(\mathcal{T}_1, \dots, \mathcal{T}_d)f$ converge a.e. (and in L_1).*

Proof: It is enough to prove only convergence of $A_n(T_1, \dots, T_d)f$, since by substituting $\mathcal{T}_i$ for T_i we'll obtain the other convergence.

L_1-convergence of $A_n(T_1, \ldots, T_d)f$ follows from applying lemma 2.3 to $\mathcal{T}_1, \ldots, \mathcal{T}_d$ and then using theorem 2.2. By proposition 1.1, each T_i is mean ergodic.

Almost everywhere convergence will be proved by induction on the number of operators. It is clear from the assumptions that any subset of $\{T_1, \ldots, T_d\}$ also satisfies the assumptions of the theorem. The case $d = 1$ is proposition 1.1.

Assume that the theorem has been proved for $d-1$ operators. We first prove a.e. convergence for f in a dense subset, namely the set obtained in lemma 2.7. So let

$$f = \sum_{i=1}^{d} R_i g_i + \Pi_{k=1}^{d}(I - T_k)h; \; h \in L_1, \; T_i g_i = g_i \in L_1$$

with $R_i = \sum_j B_{ij}$, $B_{ij} = \pm\Pi_{k=1}^{d} S_k$ with $S_k \in \{I, T_k\}$. Then $T_i g_i = g_i$ and the commutativity yield $A_n(T_1, \ldots, T_d)B_{ij}g_i = \Pi_{k\neq i}A_n(T_k)B_{ij}g_i$, which converges a.e. by the induction hypothesis, being a $(d-1)$-dimensional average. Let $V = \Pi_{k=1}^{d}\mathcal{T}_k$. Since $d \geq 2$, we obtain

$$\int \sum_{n=1}^{\infty} \frac{V^n|h|}{n^d} d\mu = \sum_{n=1}^{\infty} \frac{\|V^n|h|\|_1}{n^d} \leq \|h\|_1 \sum_{n=1}^{\infty} \frac{1}{n^d} < \infty.$$

Hence the series integrated converges a.e., and

$$|A_n(T_1, \ldots, T_d)\Pi_{k=1}^{d}(I - T_k)h| = |\frac{1}{n^d}(\Pi_{k=1}^{d} T_k)^n h| \leq \frac{V^n|h|}{n^d} \to 0 \text{ a.e.}$$

This proves the a.e. convergence on a dense subset of L_1.

By theorem 2.2(ii), the Brunel operator U is mean ergodic, and by proposition 1.1 (i.e. by Ito-Kim's theorem), $A_n(U)f$ converges a.e. for every $f \in L_1$. By Brunel's estimate [K, p.214],

$$|A_n(T_1, \ldots, T_d)f| \leq cA_{n'}(U)|f| \leq c \sup_{k\geq 1} A_k(U)|f|$$

which proves that $\sup_{n\geq 1} |A_n(T_1, \ldots, T_d)f| < \infty$ a.e. for $f \in L_1$. From Banach's principle [K, p.64] we obtain that for *every* $f \in L_1$, $A_n(T_1, \ldots, T_d)f$ converges a.e.

Remarks The proof, by induction, requires not only the mean ergodicity of U, but also mean ergodicity of the Brunel operators for subsets of $\{T_1, \ldots, T_d\}$, so mean ergodicity of $\tilde{T}_i$ (which yields that of T_i) is almost *required* for our proof. Commutativity of the moduli is used only for proving mean ergodicity of the different Brunel operators.

Acknowledgement Part of this research was carried out during the second author's visit to North Dakota State University. He wishes to thank NDSU for its warm hospitality.

References

[AS] M. A. Akcoglu and L. Sucheston, A ratio ergodic theorem for superadditive processes, Z. Wahrscheinlichkeitstheorie verw. Geb. 44 (1978), 269-278.

[B] A. Brunel, Theéorème ergodique ponctuel pour un semi-groupe commutatif finiment engendré de contractions de L_1, Ann. inst. Poincaré B9 (1973), 327-343.

[BS] A. Brunel and L. Sucheston, Sur l'existence de dominants exacts pour un processus sur-additif, CRAS Paris A-288 (1979), 153-155.

[DL] Y.Derriennic and M. Lin, On invariant measures and ergodic theorems for positive operators, J. Funct. Anal. 13 (1973), 252-267.

[I_1] Y. Ito, Uniform integrability and the pointwise ergodic theorem, Proc. Amer. Math. Soc. 16 (1965), 222-227.

[I_2] Y. Ito, Invariant measures and the pointwise ergodic theorem, Comment. Math. Univ. St. Pauli. 30 (1981), 193-201.

[K] U. Krengel, Ergodic theorems, de Gruyter, Berlin, 1985.

[Ki] C.W. Kim, A generalization of Ito's theorem concerning the pointwise ergodic theorem, Ann. Math. Stat. 39 (1968), 2145-2148.

[M] S.A. McGrath, Some ergodic theorems for commuting L_1 contractions, Studia Math. 70(1981), 153-160.

[W] H. Weiner, Invariant measures and Cesàro summability, Pacific J. 25 (1968), 621-629.

Multi–Parameter Moving Averages

By

Roger L. Jones[1]
Department of Mathematics
DePaul University
2219 N. Kenmore
Chicago, IL 60614

and

James Olsen[2]
Department of Mathematics
North Dakota State University
Fargo, North Dakota 58105

Introduction. Consider a dynamical system, (X,Σ,m,T) where (X,Σ,m) is a probability space and T is a measure preserving point transformation from X onto itself. In this setting Akcoglu and del Junco [1] considered averages of the form

$$A_k f(x) = \frac{1}{\sqrt{k}} \sum_{j=1}^{[\sqrt{k}]} f(T^{k+j}x).$$

They showed that these averages can diverge a.e. for some $f \in L^\infty(X)$. More generally, let $\{(n_k,\ell_k)\}_{k=1}^\infty$ denote a sequence of points in $\mathbb{Z}\times\mathbb{Z}^+$. In [3] and [4] averages of the form

$$A_k f(x) = \frac{1}{\ell_k} \sum_{j=1}^{\ell_k} f(T^{n_k+j}x)$$

are considered, and necessary and sufficient conditions on $\{(n_k,\ell_k)\}_{k=1}^\infty$ are given to insure almost everywhere convergence for all $f \in L^1(X)$. Further, when this fails, it is shown that given $\epsilon>0$, there exists a set B with $m(B) < \epsilon$, such that $\limsup A_k \chi_B(x) = 1$ a.e. and $\liminf A_k \chi_B(x) = 0$ a.e. The positive results for $f \in L^p(X)$, $p>1$ were generalized in [7] from the case of measure preserving point transformations to the case of

1. Partially supported by NSF grant DMS–8910947
2. Much of the work presented here was done while the author was visiting Northwestern University, whose support and hospitality he gratefully acknowledges.

ISBN 0-12-085520-8

T a Dunford–Schwartz operator, that is to the case of linear operators that are simultaneously contractions on $L^1(X)$ and $L^\infty(X)$. In the present paper we extend these results to the case of several transformations.

For $A \subset \mathbb{Z}^d$ we will use $|A|$ to denote the number of lattice points in the set A. We will be interested both in functions on $\mathbb{Z}^d$ and functions on X. To help avoid confusion regarding the domain, we will use φ to denote a function on $\mathbb{Z}^d$ and f or g to denote a function on X.

Let $\mathbf{T} = (T_1,T_2,\cdots,T_d)$ be a vector of d transformations from a measure space (X,Σ,m) onto itself. For $\mathbf{j} = (j_1,\cdots,j_d)$, define $\mathbf{T}^{\mathbf{j}}(f)(x) = T_1^{j_1}T_2^{j_2}\cdots T_d^{j_d}(f)(x)$. Let $\{R_k\}_{k=1}^{\infty}$ denote a sequence of rectangles in $\mathbb{Z}^d$ with sides parallel to the axis. Consider averages of the form

$$A_k f(x) = \frac{1}{|R_k|} \sum_{\mathbf{j} \in R_k} \mathbf{T}^{\mathbf{j}}(f)(x).$$

We will be interested on conditions on the sequence $\{R_k\}_{k=1}^{\infty}$ and the multi–parameter transformation $\mathbf{T}$ to insure almost everywhere convergence for all $f \in L^p(X)$ for some p, $1 \leq p \leq \infty$.

<u>Definition:</u> We will say that τ is a <u>non–singular point transformation</u> if $\tau:X \to X$, and if for each $A \in \Sigma$ $\tau^{-1}(A)$ and $\tau(A)$ are in Σ and $m(\tau^{-1}A) > 0$ if and only if $m(A)>0$.

<u>Definition:</u> The transformation $\mathbf{T}$ is <u>induced by non–singular point transformations</u> if for each k, $k = 1,2,\cdots,d$, we have $T_k(f)(x) = w_k(x)f(\tau_k x)$, and each τ_k is a non–singular point transformation.

<u>Remark.</u> If T_k is to be a positive invertible isometry of $L^p(X)$, p fixed, $1 < p < \infty$, then w_k is given by $w_k(x) = \left(\frac{d\mu \circ \tau_k}{d\mu}\right)^{1/p}(x)$.

When T is induced by non–singular point transformations we will often write Tx instead of τx.

<u>Definition:</u> The transformation T is <u>non–periodic</u> if T is induced by non–singular point transformations, and if for every d–tupple of integers, $\mathbf{j} = (j_1, j_2, \cdots, j_d) \neq (0,0,\cdots,0)$, we have

$$m\{x: T^{\mathbf{j}}(x) = x\} = 0.$$

<u>Definition:</u> The transformation T is <u>induced by measure preserving point transformations</u> if for each k, $T_k(f)(x) = f(\tau_k x)$ where τ_k is a measure preserving point transformation. If each τ_k is also ergodic we will say T is <u>induced by ergodic measure preserving point transformations</u>.

<u>Section I: Averages over Cubes.</u>

We will first prove a d–transformation analog of Theorem 1 from [3]. Let $\{Q_k\}_{k=1}^{\infty}$ denote a sequence of cubes in $\mathbb{Z}^d$, with sides parallel to the axis. Let $\mathbf{n}_k = (n_k^1, n_k^2, \cdots, n_k^d)$ denote the coordinate of the lower left corner of Q_k and let ℓ_k denote the side length. We associate with the sequence of sets, $\{Q_k\}_{k=1}^{\infty}$ a subset Ω of $\mathbb{Z}^d \times \mathbb{Z}^+$ where $\Omega = \{(\mathbf{n}_k, \ell_k)\}_{k=1}^{\infty}$. For any $\alpha > 0$ we can define

$$\Omega_\alpha = \{(\mathbf{z}, s): \mathbf{z} \in \mathbb{Z}^d, \ |\mathbf{z} - \mathbf{n}_k| \le \alpha(s - \ell_k) \text{ for some } (\mathbf{n}_k, \ell_k) \in \Omega\}$$

and the cross section of Ω_α at integer height λ by

$$\Omega_\alpha(\lambda) = \{\mathbf{n}: (\mathbf{n}, \lambda) \in \Omega_\alpha\}.$$

Define the maximal operator

$$M_\Omega f(x) = \sup_k \left| \frac{1}{|Q_k|} \sum_{\mathbf{j} \in Q_k} T^{\mathbf{j}}(f)(x) \right|.$$

We then have:

Theorem 1.1 *Let T be induced by commuting ergodic measure preserving point transformations.*

a) Assume for some $\alpha>0$ there exists a constant $A < \infty$ such that $|\Omega_\alpha(\lambda)| < A\lambda^d$ for each integer $\lambda > 0$, then $M_\Omega f$ is weak type (1,1) and strong type (p,p) for each $p>1$.

b) If M_Ω is weak type (p,p) for some finite $p>0$, then for every $\alpha>0$, there exists $A_\alpha<\infty$ such that for all integer $\lambda > 0$, we have $|\Omega_\alpha(\lambda)| \leq A_\alpha\lambda^d$.

Remark. This is the ergodic theory analog of the real variable result of Nagel and Stein [10], and the proof is strongly related to the proof of their result by Suiro [15].

Before we begin the proof of the theorem, we state and prove a version of the Calderón transfer principle [5] as it relates to our problem. Note that the fact that the transformations commute is used in an essential way in the proof.

Lemma 1.1 (The Calderón Transfer Principle) *Let Ω and $M_\Omega f$ be defined as above. Let $\varphi \in \ell^p(\mathbb{Z}^d)$ and define*

$$M_\Omega\varphi(s) = \sup_{k<N}\left|\frac{1}{|Q_k|}\sum_{j\in Q_k}\varphi(s+j)\right|.$$

If for some p, $1 \leq p < \infty$, $M_\Omega\varphi$ is weak type (p,p), then $M_\Omega f$ is also weak type (p,p) for that same p, and with the same weak type constant.

Proof. We will give the details of the proof only in the special case of two transformations. The more general case follows by the same argument.

Let (n_k^1,n_k^2) denote the lower left corner of the rectangle Q_k, and assume the side length of $Q_k = \ell_k$. Assume that T_1 and T_2 are induced by the measure preserving point transformations τ_1 and τ_2 respectively. Let

$$A_k f(x) = \frac{1}{\ell_k^2} \sum_{i=0}^{\ell_k-1} \sum_{j=0}^{\ell_k-1} f(\tau_1^{n_k^1+i} \tau_2^{n_k^2+j} x),$$

and $M_N f(x) = \sup_{k \in I_N} |A_k f(x)|$ where

$$I_N = \{k: n_k^1+\ell_k \leq N, n_k^2+\ell_k \leq N\}.$$

Let $E = \{x: M_N f(x) > \lambda\}$. We will show that $m(E) \leq \frac{c}{\lambda^p} \|f\|_p^p$ where c is independent of N. Letting $N \to \infty$ will complete the proof. Fix a very large integer L. For each x, define

$$\varphi_x(i,j) = f(\tau_1^i \tau_2^j x) \chi_{[0,L+N]}(i)\, \chi_{[0,L+N]}(j).$$

Define the averaging operators on $\ell^p(\mathbb{Z}\times\mathbb{Z})$ by

$$A_k \varphi(r,s) = \frac{1}{\ell_k^2} \sum_{i=0}^{\ell_k-1} \sum_{j=0}^{\ell_k-1} \varphi(r+n_k^1+i, s+n_k^2+j),$$

and the maximal operator $M_N \varphi(r,s) = \sup_{k \in I_N} |A_k \varphi(r,s)|$. Note that for $0 \leq r,s \leq L$, we have

$$A_k \varphi_x(r,s) = \frac{1}{\ell_k^2} \sum_{i=0}^{\ell_k-1} \sum_{j=0}^{\ell_k-1} \varphi_x(r+n_k^1+i, s+n_k^2+j)$$

$$= \frac{1}{\ell_k^2} \sum_{i=0}^{\ell_k-1} \sum_{j=0}^{\ell_k-1} f(\tau_1^{r+n_k^1+i} \tau_2^{s+n_k^2+j} x)$$

$$= \frac{1}{\ell_k^2} \sum_{i=0}^{\ell_k-1} \sum_{j=0}^{\ell_k-1} f(\tau_1^{n_k^1+i} \tau_2^{n_k^2+j} \tau_1^r \tau_2^s x)$$

$$= A_k f(\tau_1^r \tau_2^s x).$$

Consequently, for $0 \leq r,s \leq L$, we have $\{M_N f(\tau_1^r \tau_2^s x) > \lambda\} = \{M_N \varphi_x(r,s) > \lambda\}$. From this we see that

$$m(E) = \int_X \chi_E(x)\, dm(x)$$

$$= \frac{1}{(L+1)^2} \sum_{r=0}^{L} \sum_{s=0}^{L} \int_X \chi_E(\tau_1^r \tau_2^s x)\, dm(x)$$

$$= \frac{1}{(L+1)^2} \int_X \sum_{r=0}^{L} \sum_{s=0}^{L} \chi_E(\tau_1^r \tau_2^s x)\, dm(x)$$

$$= \frac{1}{(L+1)^2} \int_X |\{(r,s): 0 \leq r,s \leq L,\ M_N \varphi_x(r,s) > \lambda\}|\, dm(x)$$

$$\leq \frac{1}{(L+1)^2} \int_X \frac{c}{\lambda^p} \|\varphi_x\|_p^p\, dm(x)$$

$$= \frac{1}{(L+1)^2} \int_X \frac{c}{\lambda^p} \sum_{r=0}^{L+N} \sum_{s=0}^{L+N} |f(\tau_1^r \tau_2^s x)|^p\, dm(x)$$

$$\leq \frac{1}{(L+1)^2} \frac{c}{\lambda^p} \sum_{r=0}^{L+N} \sum_{s=0}^{L+N} \int_X |f(\tau_1^r \tau_2^s x)|^p\, dm(x)$$

$$\leq \frac{1}{(L+1)^2} \frac{c}{\lambda^p} (L+N+1)^2 \int_X |f(x)|^p\, dm(x).$$

Now let $L \to \infty$ to obtain the desired inequality. □

__Remark:__ In the above argument we showed that $\{M_N f(\tau_1^r \tau_2^s x) > \lambda\} = \{M_N \varphi_x(r,s) > \lambda\}$. To do this we used the fact that $f(\tau_1^{r+n_k^1+i} \tau_2^{s+n_k^2+j} x) = f(\tau_1^{n_k^1+i} \tau_2^{n_k^2+j} \tau_1^r \tau_2^s x)$, which is true in general only if τ_1 and τ_2 commute. At this point we do not know how to prove the weak type (1,1) inequality if τ_1 and τ_2 do not commute. Further we used the fact that

$$\int_X \chi_E(x)\, dm(x) = \int_X T_1^r T_2^s \chi_E(x)\, dm(x)$$

and that $T_1^r T_2^s \chi_E(x)$ is still a characteristic function. While these two restrictions can be relaxed slightly (see [2]) it is still unclear exactly what hypothesis are required on the transformations $T_1, \cdots, T_d$ which will allow us to transfer a weak type (p,p) inequality. However, with the above proof, the requirement that $T_1, \cdots, T_d$ commute seems essential.

<u>Proof of Theorem 1.1, part a).</u> In the context of the maximal function problem with commuting transformations, Lemma 1.1, above allows us to transfer our considerations from (X,Σ,m,T) to the integer lattice $\mathbb{Z}^d$, with translation. Fix a point $x \in X$, and a large integer N.

Define the non centered Hardy–Littlewood maximal function,

$$\varphi^\star(p) = \sup_{Q \ni p} \left| \frac{1}{|Q|} \sum_{j \in Q} \varphi(j) \right|$$

where the sup is taken over all cubes containing the point p. We can with out loss of generality assume that $\varphi \geq 0$. Decompose the domain of φ, which is just $\mathbb{Z}^d$, into a collection of disjoint cubes, $\{B_j\}$ where $\varphi^\star(p) > \lambda$ for each $p \in \bigcup_{j=1}^{\infty} B_j$. In the complement of the union of the cubes we have $\varphi^\star(p) \leq \lambda$. Assume $B_j = B_j(x_j,s_j)$ is centered at x_j and has side length s_j. (We will take the side length to be the number of lattice points on a side.) We can also assume that if $B_j^\star$ is B_j with each side expanded by a factor of 4, then $B_j^\star$ contains points where $\varphi^\star < \lambda$. (See Stein [13] page 16 for this decomposition in the case of $\mathbb{R}^d$. We can think of $\varphi^\star(x)$ as a function on $\mathbb{R}^d$, use the decomposition, and restrict back to $\mathbb{Z}^d$.) Note that $s_j \geq 1$ for all j.

Now let $M_\Omega \varphi(p) > 4^d\lambda$, then there exists a point $(n,\ell) \in \Omega$ such that for Q the cube with left hand corner at n and side length ℓ, we have

$$\frac{1}{|Q|} \sum_{j \in Q} \varphi(p+j) > 4^d\lambda.$$

Thus the point $p+n \in \{\varphi^\star > \lambda\}$, and hence $p + n \in B_j$ for some integer j. We now show that the cube $p + Q$ is entirely contained in $B^{\star}_j$. There are two cases. Case 1. Assume $\ell \leq s_j$. Then the containment is obvious. Case 2. Assume $\ell > s_j$. Associate with the cube $p + Q$ a new cube $\tilde{Q}$ which has the same center as $p + Q$, and each side is 4 times as large. Then $\tilde{Q}$ contains B_j. However, no point of $\tilde{Q}$ can belong to the set where $\varphi^{\star} \leq \lambda$. To see this, note that if $\tilde{Q}$ were to contain such a point, then

$$\frac{1}{|\tilde{Q}|}\sum_{j\in\tilde{Q}}\varphi(j) \leq \lambda,$$

but we know

$$\frac{1}{|\tilde{Q}|}\sum_{j\in\tilde{Q}}\varphi(j) \geq \frac{1}{|\tilde{Q}|}\sum_{j\in Q}\varphi(j)$$

$$\geq \frac{|Q|}{|\tilde{Q}|}\frac{1}{|Q|}\sum_{j\in Q}\varphi(j) > \frac{|Q|}{|\tilde{Q}|}4^d\lambda \geq \lambda,$$

a contradiction. However $B^{\star}$ does contain points where $\varphi^{\star} \leq \lambda$. From this we know that the cube Q does not contain $B^{\star}$. Thus $\ell < 4s_j$. In either case we know that $\ell \leq 4s_j$.

Let $c = 4[1+\frac{1}{\alpha}]+1$. For each j, define $\tilde{\mathbf{x}}_j$ to be the lattice point closest to $\mathbf{x}_j$. (If more than one lattice point is the same distance away, pick the any one of them.) Then we have

$$\begin{aligned}
|(\tilde{\mathbf{x}}_j-\mathbf{p}) - \mathbf{n}| &\leq |\mathbf{x}_j - (\mathbf{p}+\mathbf{n}) + (\tilde{\mathbf{x}}_j - \mathbf{x}_j)| \\
&\leq |\mathbf{x}_j - (\mathbf{p}+\mathbf{n})| + |\tilde{\mathbf{x}}_j - \mathbf{x}_j| \leq s_j + 1 \leq 2s_j \text{ (since } p+n \in B_j) \\
&\leq \alpha(4s_j + \tfrac{1}{\alpha}4s_j - 4s_j) \\
&\leq \alpha(cs_j - 4s_j) \\
&\leq \alpha(cs_j - \ell) \text{ (since } \ell < 4s_j).
\end{aligned}$$

Therefore, by definition, $(\tilde{\mathbf{x}}_j - \mathbf{p}, cs_j)$ is contained in Ω_α, or $\tilde{\mathbf{x}}_j - \mathbf{p} \in \Omega_\alpha(cs_j)$. This implies that $\mathbf{p} \in \tilde{\mathbf{x}}_j - \Omega_\alpha(cs_j)$. Using this we see that

$$\{p: M_\Omega\varphi(p) > 4(2^d)\lambda\} \subset \bigcup_j \{\tilde{\mathbf{x}}_j - \Omega_\alpha(cs_j)\}.$$

Taking measures (the counting measure on $\mathbb{Z}^d$), we have

$$
\begin{aligned}
|\{p\colon M_\Omega\varphi(p) > 4^d\lambda\}| &\le \sum_j |\tilde{x}_j - \Omega_\alpha(cs_j)| \\
&\le \sum_j |\Omega_\alpha(cs_j)| \\
&\le \sum_j A|cs_j|^d \\
&\le Ac^d \sum_j |B_j| \le Ac^d|\cup B_j| \\
&\le A\frac{C}{\lambda}\|\varphi\|_{\ell^1},
\end{aligned}
$$

where the last inequality uses the fact that $\cup B_j = \{\varphi^\star > \lambda\}$, and the fact that the Hardy–Littlewood maximal function is weak type (1,1).

From this and the transfer principle, we obtain the weak type (1,1) inequality. Using the fact that we have strong type (∞,∞), and the Marcinkiewicz interpolation theorem, we see that the operator is strong type (p,p) for each $p>1$. □

<u>Proof of Theorem 1.1, part b).</u> Assume that M_Ω is weak type (p,p). Then there exists a constant c such that for each choice of $\lambda>0$ and f, we have $|\{M_\Omega f \ge \lambda\}| \le c\lambda^{-p}\|f\|_p^p$. Fix λ and choose a large integer N such that $\Omega_\alpha(\lambda) \subset [0,N]^d$. Form a d–parameter Kakutani–Rohlin tower of side length N, and error less than 1/2. In other words, we find a set B such that each of the sets $T^{\mathbf{m}}B$ are disjoint, for $\mathbf{m} = (m_1,\cdots,m_d)$, with $0 \le m_i \le N-1$, $i=1,2,\cdots,d$, and $\mu(\bigcup_{\mathbf{m}} T^{\mathbf{m}}B) > 1/2$. (See [8] or [9] for the details of this decomposition. Also see Olsen, [11].) Let $(\mathbf{z},\lambda)$ be in Ω_α. Then by definition of Ω_α we know $|\mathbf{z}-\mathbf{n}_k| \le \alpha(\lambda-\ell_k)$ for some point $(\mathbf{n}_k,\ell_k) \in \Omega$. Note that this implies $|\mathbf{z}-\mathbf{n}_k| \le \alpha\lambda$

and $\lambda \geq \ell_k$. Let $m = [(\alpha+1)\lambda] + 1$. Let Q be a cube with side length $2m$ and center at $c = (N-m,\cdots,N-m)$. Define a function f to be λ on $\bigcup_{i \in Q} T^i B$ and zero elsewhere. Let x by a point in the set $T^c B$. Thus

$$M_\Omega f(\tau^{-z} x) \geq \frac{1}{|Q_k|} \sum_{j \in Q_k} f(\tau^{-z+j} x).$$

Since $|-z+n_k| < \alpha\lambda \leq m$, the sum starts in the support of f. The sum also ends in the support of f because $|-z+n_k+j| \leq |-z+n_k|+\ell_k \leq \alpha\lambda + \lambda \leq (\alpha+1)\lambda \leq m$. Thus the entire sum is within the support of f, and we conclude that $M_\Omega f(\tau^{-z} x) \geq \lambda$. This is true for all z such that $(z,\lambda) \in \Omega_\alpha$, i.e. for all z in $\Omega_\alpha(\lambda)$. Thus $|\{M_\Omega f > \lambda\}| \geq |\Omega_\alpha(\lambda)|/(2N^d)$. On the other hand, we know that $\|f\|_p^p = \lambda^p (2m)^d \mu(B)$. However, $M_\Omega f$ is weak type (p,p) and hence we have $|\{M_\Omega f > \lambda\}| \leq \frac{c}{\lambda^p}\|f\|_p^p \leq \frac{c}{\lambda^p} \lambda^p (2m)^d \mu(B)$. Thus $|\Omega_\alpha(\lambda)|/(2N^d) \leq c(2m)^d \mu(B) \leq (2[(\alpha+1)\lambda])^d/(N^d/2)$. This becomes $|\Omega_\alpha(\lambda)| \leq < A\lambda^d$ for a suitable choice of A. □

Unfortunately it can sometimes be difficult to check directly if Ω_α satisfies the growth condition required to apply Theorem 1.1. However there is an equivalent condition that can be more easily checked.

<u>**Theorem 1.2.**</u> *The set Ω_α satisfies the growth condition $|\Omega_\alpha(\lambda)| < c\lambda^d$ if and only if for each i, $1 \leq i \leq d$,*

$$\Omega_\alpha^i \equiv \{(z,s) : |z - n_k^i| < \alpha(s - \ell_k)\}$$

satisfies the linear growth condition $|\Omega_\alpha^i(\lambda)| < c_i\lambda$.

<u>Proof.</u> First assume that Ω_α^i satisfies the growth condition $|\Omega_\alpha^i(\lambda)| < c_i\lambda$ for each i. We will then show $|\Omega_\alpha(\lambda)| < c\lambda^d$. This follows since

$$\begin{aligned}
|\Omega_\alpha(\lambda)| &= |\{z\colon |z-n_k| < \alpha(\lambda-\ell_k) \text{ for some } k\}| \\
&\le |\{z\colon |z_i - n_k^i| < \alpha(\lambda-\ell_k) \text{ for each } i \text{ for some } k\}| \\
&\le \prod_{i=1}^{d} |\{z_i\colon |z_i - n_k^i| < \alpha(\lambda-\ell_k) \text{ for some } k\}| \\
&\le \prod_{i=1}^{d} c_i\lambda = c\lambda^d,
\end{aligned}$$

where $c = \prod_{i=1}^{d} c_i$.

To see that the "only if" direction holds, assume that for some i, Ω_α^i fails to satisfy the growth condition $|\Omega_\alpha^i(\lambda)| < c_i\lambda$ for any finite choice of c_i. With no loss of generality, we can assume that $i = 1$. Thus there is an increasing sequence $\{\lambda_n\}_{n=1}^{\infty}$ such that $|\Omega_\alpha^1(\lambda_n)| > n\lambda_n$.

For each value of n, consider the region $\Omega_\alpha(\lambda_n)$. The measure of the projection of this region onto the first coordinate axis is at least $n\lambda_n$. Place a cone of aperture α at each point of $\Omega_\alpha(\lambda_n)$. Call the resulting region $\tilde{\Omega}_{\alpha,n}$ and denote its cross sectional area at height λ by $\tilde{\Omega}_{\alpha,n}(\lambda)$. Clearly $\tilde{\Omega}_{\alpha,n} \subset \Omega_\alpha$, and

$$\begin{aligned}
|\tilde{\Omega}_{\alpha,n}(2\lambda_n)| &\ge c(2\lambda_n - \lambda_n)^{d-1} |\Omega_\alpha^1(\lambda_n)| \\
&\ge c\,\lambda_n^{d-1} n\lambda_n \\
&= cn\lambda_n^d,
\end{aligned}$$

where c depends only on α. Consequently,

$$|\Omega_\alpha(\lambda_n)| \ge c\,n\,\lambda_n^d$$

which implies Ω_α does not satisfy the required growth condition. $\square$

Using the maximal inequality from Theorem 1.1 and Theorem 1.2, we can now easily prove an almost everywhere convergence theorem.

Theorem 1.2. *If M_Ω is weak type (1,1), and T is induced by commuting measure preserving point transformations, then the averages $A_k f(x)$ converge a.e. for all $f \in L^p(X)$, $1 \leq p < \infty$.*

Proof. We have assumed that the maximal function is weak type (1,1), and consequently weak type (p,p) for all p, $1 \leq p < \infty$. Thus it will be enough to establish convergence on a dense class. To keep the notation from obscuring the idea, we give the proof in the case d = 2. First consider the class

$$G_2 = \{g \colon g = h_2 + (g_2 - T_2 g_2), \text{ with } h_2 \in L^\infty(X) \text{ invariant under } T_2, \text{ and } g_2 \in L^\infty(X)\}$$

This class is dense. Now consider the class

$$G_1 = \{g_1 \colon g_1 = h_1 + (g - T_1 g), \text{ with } h_1 \in L^\infty(X) \text{ invariant under } T_1, \ g \in G_2\}.$$

This class is also dense in each $L^p(X)$. Let $f \in G_1$. Then

$$f = h_1 + (h_2 + g_2 - T_2 g_2) - T_1(h_2 + g_2 - T_2 g_2)$$

For such functions we have

$$\frac{1}{\ell_k^2} \sum_{i=n_k^i}^{n_k^i+\ell} \ \sum_{j=n_k^2}^{n_k^2+\ell_k} T_1^i T_2^j f = \frac{1}{\ell_k^2} \sum_{i=n_k^1}^{n_k^1+\ell_k} \ \sum_{j=n_k^2}^{n_k^2+\ell_k} T_1^i T_2^j h_2$$

$$+ \frac{1}{\ell_k^2} \sum_{i=n_k^2}^{n_k^2+\ell_k} T_2^i \Big(\sum_{j=n_k^1}^{n_k^1+\ell_k} T_1^j ((h_2 + g_2 - T_2 g_2) - T_1(h_2 + g_2 - T_2 g_2)) \Big)$$

$$= \frac{1}{\ell_k^2} \sum_{i=n_k^1}^{n_k^1+\ell_k} \ \sum_{j=n_k^2}^{n_k^2+\ell_k} T_1^i T_2^j h_2$$

$$+ \frac{1}{\ell_k^2} \sum_{i=n_k^2}^{n_k^2+\ell_k} T_2^i \Big(T_1^{n_k^1} (h_2 + g_2 - T_2 g_2) - T_1^{n_k^1+\ell_k+1} (h_2 + g_2 - T_2 g_2) \Big)$$

$$= \frac{1}{\ell_k} \sum_{i=n_k^1}^{n_k^1+\ell_k} T_1^i h_2 + \frac{1}{\ell_k}(T_1^{n_k^1} h_2 - T_1^{n_k^1+\ell_k+1} h_2) + \frac{1}{\ell_k} T_1^{n_k^1} (T_2^{n_k^2} g_2 - T_2^{n_k^2+\ell_k+1} g_2)$$

$$- \frac{1}{\ell_k} T_1^{n_k^1+\ell_k+1} (T_2^{n_k^2} g_2 - T_2^{n_k^2+\ell_k+1} g_2)$$

$$= I + II + III + IV.$$

We must show that each of these terms converge. To see that I) converges note that by Theorem 1.2, $(n_k^1, \ell_k)_{k=1}^{\infty}$ satisfies the growth condition necessary to apply the one parameter results from [3]. To see that II), III) and IV) converge, note that both T_1 and T_2 are contractions on $L^{\infty}(X)$, g_2 and h_2 are in $L^{\infty}(X)$, and ℓ_k goes to infinity, thus we have convergence to zero. Additional commuting transformations can be handled in the same way. □

Using Theorem 1.2 it is now easy to construct examples of both "good" regions, i.e. regions where the averages converge a.e. for all $f \in L^1(X)$, and "bad" regions, i.e. regions where divergence can occur even for $f \in L^{\infty}(X)$.

Example 1. Let $n_k^i = 2^{2^k}$ for each i, $i = 1,2, \cdots, d$ and $\ell_k = \sqrt{2^{2^k}}$. Then each Ω_α^i satisfies the growth condition (see [3]) and consequently by Theorem 1.2 so does Ω_α. Consequently by Theorems 1.1 and 1.3, we have convergence a.e. for d commuting measure preserving transformations.

Example 2. Let $n_k^1 = k^2$ and $\ell_k = k$, then for any choice of n_k^i, $i = 2,\cdots,d$, we have that Ω_α^1 fails to satisfy the growth condition, hence by Theorem 1.2, Ω_α fails to satisfy the growth condition $|\Omega_\alpha(\lambda)| < c\lambda^d$. Consequently $M_\Omega f$ fails to be weak type (p,p) for any p, $1 \leq p < \infty$.

Section 2: Averages over Rectangles.

Theorem 1.1 and Theorem 1.3 can be partially generalized in several ways. Theorem 2.1 below generalizes Theorems 1.1 and 1.3 both by allowing more general operators and more general regions. Before stating the theorem, we give some definitions.

Definition: A linear operator $T: L^p(X) \to L^p(X)$ is called Lamperti if whenever f and g have disjoint supports, so do Tf and Tg.

Let $\{R_k\}_{k=1}^{\infty}$ be a sequence of rectangles with lower left corner at $\mathbf{n}_k$ and side lengths $\boldsymbol{\ell}_k$. As before $\mathbf{n}_k$ denotes $(n_k^1, \cdots n_k^d)$ and let $\boldsymbol{\ell}_k = (\ell_k^1, \cdots, \ell_k^d)$.

Definition: For $i = 1,2,\cdots,d$ the set $\Omega^i = \{(n_k^i, \ell_k^i)\}_{k=1}^{\infty}$ is the projection of $\{\mathbf{n}_k, \boldsymbol{\ell}_k\}_{k=1}^{\infty}$ onto the i th coordinate axis,

$$\Omega_\alpha^i = \{(z,s) \in \mathbb{Z}^d \times \mathbb{Z}^+ : |z - n_k^i| \leq \alpha(s - \ell_k^i) \text{ for some } k\}.$$

and

$$\Omega_\alpha^i(\lambda) = \{z: (z,\lambda) \in \Omega_\alpha^i\}.$$

With this notation, we now have:

Theorem 2.1. *Let $T_1, \cdots, T_d$ be power bounded Lamperti operators, or operators bounded by positive contractions of L^p, p fixed, $1 < p < \infty$. Assume the sequence of rectangles, $\{R_k\}_{k=1}^{\infty}$ satisfy for each $i = 1, 2, \cdots, d$, the linear growth condition $\Omega_1^i(\lambda) \leq c_i \lambda$ and $\ell_{k+1}^i > \ell_k^i$ for all k. Define*

$$A_k f(x) = \frac{1}{\ell_k^d} \sum_{j_1=n_k^1}^{n_k^1+\ell_k-1} \sum_{j_2=n_k^2}^{n_k^2+\ell_k-1} \cdots \sum_{j_d=n_k^d}^{n_k^d+\ell_k-1} T_1^{j_1} T_2^{j_2} \cdots T_d^{j_d}(f)(x)$$

then

a) $Mf(x) = \sup_k |A_k f(x)|$ *is a bounded operator from* $L^p(X)$ *to* $L^p(X)$.

b) $\lim_{k\to\infty} A_k f(x)$ *exists a.e.*

Remark 1. It is clear that the restrictions on $\{(\mathbf{n}_k, \boldsymbol{\ell}_k)\}$ to have convergence for all $f \in L^1(X)$ must be different from the restrictions necessary for convergence for all $f \in L^p(X)$. To see this, simply note that if we take all $\mathbf{n}_k = (0,0,\cdots,0)$, and $\{\ell_k\}$ to include the entire integer lattice in the first quadrant, with the restriction that for $k>j$, the distance from ℓ_k to $(0,0,\cdots,0)$ is greater than or equal to the distance from ℓ_j to $(0,0,\cdots,0)$. It is now easy to see that the associated maximal operator is not weak type $(1,1)$, but is strong type (p,p) for each $p>1$. (See Zygmund [16].)

Remark 2. The extra condition $\ell_{k+1}^i \geq \ell_k^i$ is not required if we know that T_i is a power bounded on $L^\infty(X)$. In general it was needed in [7] to show that the one–parameter averages converge a.e. on a dense subset of $L^p(X)$. In the case of T_i power bounded on $L^\infty(X)$ the convergence on a dense set can be proven directly.

Proof of Theorem 2.1. Part a) follows easily by induction on the number of transformations. Define for each i, the operator

$$A_k^i f(x) = \frac{1}{\ell_k} \sum_{j_i=n_k^i}^{n_k^i+\ell_k-1} T_i^{j_i}(f)(x).$$

Note that for each i, $\Omega_1^i(\lambda)$ satisfies the linear growth condition, and consequently for each i, the operator $M_i f(x) = \sup_k |A_k^i f(x)|$ is a bounded operator from $L^p(X)$ to $L^p(X)$. This follows from Theorem 1 in [3] in the case T_i is induced by a measure preserving point transformation, and from [7] in the more general case. Consequently from this and the fact that $Mf(x) \le M_1(M_2 \cdots (M_d f(x)) \cdots)$, a) follows. □

Proof of Theorem 2.1 part b. The usual path to prove b) is to establish convergence on a dense set and then apply the results of part a). The following argument proves convergence for all of $L^p(X)$ without the need to find a special dense class. To prove this generalization we will use a result of Sucheston [14] for deducing multi–parameter ergodic theorems from single–parameter ergodic theorems. (See also Frangos and Sucheston [6].) Sucheston's result has been specialized in [12] to the case we are interested in. For completeness, we will give a proof of this result in the case of two transformations. The reader can easily fill in the details for more transformations or can consult one of the papers mentioned above, especially [14]. The argument below is from Sucheston [14].

Let

$$A_k^i f(x) = \frac{1}{\ell_k^i} \sum_{j_i = n_k^i}^{n_k^i + \ell_k^i - 1} T_i^{j_i}(f)(x).$$

Consider the "unrestricted" limit operator,

$$\lim_{\min(r,s) \to \infty} A_r^1(A_s^2(f))(x).$$

If we could show that this limit exists a.e., then in particular we have

$$\lim_{r \to \infty} A_r^1(A_r^2 f(x)) = \frac{1}{\ell_r^2} \sum_{j_1 = n_r^1}^{n_r^1 + \ell_r^1 - 1} \sum_{j_2 = n_r^2}^{n_r^2 + \ell_r^2 - 1} T_1^{j_1} T_2^{j_2}(f)(x)$$

exists a.e. Note that by the hypothesis of our theorem, and the 1–dimensional results from [7], we have $\lim_{k\to\infty} A_k^i f(x)$ exists a.e. Denote this limit by $A_\infty^i f(x)$.

To study

$$\lim_{\min(r,s)\to\infty} A_r^1(A_s^2(f))(x)),$$

let

$$S_n(f)(x) = \sup_{s\geq n} |A_s^2(f)(x) - A_\infty^2(f)(x)|.$$

Note that the conditions on Ω_α^2 imply that $S_n(f)(x) \to 0$ for a.e. x, and $\|S_n(f)\|_p \to 0$ as $n \to \infty$. Thus for each fixed n, we have

$$\begin{aligned}\limsup_{r,s} |A_r^1 A_s^2 f(x) - A_\infty^1 A_\infty^2 f(x)| &\leq \limsup |A_r^1(A_s^2 f(x) - A_\infty^2 f(x)| \\ &\quad + \limsup |A_r^1 A_\infty^2 f(x) - A_\infty^1 A_\infty^2 f(x)| \\ &\leq \limsup |A_r^1(S_n f(x))| \\ &= A_\infty^1 S_n f(x).\end{aligned}$$

Therefore

$$\limsup_{r,s} |A_r^1 A_s^2 f(x) - A_\infty^1 A_\infty^2 f(x)| \leq A_\infty^1 S_n f(x)$$

Taking L^p norms, we have

$$\begin{aligned}\| \limsup_{r,s} |A_r^1 A_s^2(f) - A_\infty^1 A_\infty^2(f)| \|_p &\leq \| A_\infty^1 S_n(f)\|_p \\ &\leq c\| S_n(f)\|_p.\end{aligned}$$

Let $n\to\infty$, and using the fact that $\|S_n(f)\|_p \to 0$, we have

$$\| \limsup_{r,s} |A_r^1 A_s^2(f) - A_\infty^1 A_\infty^2(f)| \|_p = 0.$$

Consequently,

$$\lim_{\min(r,s)\to\infty} A_r^1 A_s^2(f)(x) = A_\infty^1 A_\infty^2(f)(x) \text{ a.e. } \square$$

Section 3: Strong Sweeping Out.

In this section, we will assume we are working with transformations that are induced by commuting ergodic measure preserving point transformations.

Definition: A family of operators, $\{T(k)\}_{k=1}^{\infty}$ is said to have the "strong sweeping out property" if given $\epsilon>0$, there exists a set B, with $\mu(B)<\epsilon$ such that

$$\limsup_{k} T(k)\chi_B = 1 \text{ a..e.}$$

$$\liminf_{k} T(k)\chi_B = 0 \text{ a..e.}$$

This property was studied in [3] where the following theorem is proved:

Theorem 3.1. *Let (X,Σ,μ) denote a probability space. Assume that $\{T(k)\}_{k=1}^{\infty}$ is a sequence of linear operators, $T(k): L^1(X) \to L^1(X)$, with the properties that*

i) $T(k) \geq 0$.

ii) $T(k)1 = 1$.

iii) The $T(k)$'s commute with a family $\{S_\alpha\}$ which is a mixing family of measure preserving transformations in the sense that for each pair of sets $A, B \in \Sigma$, and each $\rho>1$, there exists S_α in the family with the property that $m(A \cap S_\alpha^{-1}B) < \rho m(A)m(B)$.

For each n define $M_n f = \sup_{k\geq n} |T(k)f|$. Assume that

(*) For each $\epsilon>0$ and $n\in N$, there exists a sequence of sets $\{A_p\}$ such that if $E_p = \{M_n\chi_{A_p} > 1-\epsilon\}$ then $\sup_p \frac{\mu(A_p)}{\mu(E_p)} = \infty$,

then the strong sweeping out property holds.

Using this it is not hard to see that if the cone condition fails in any variable, then the averaging operators $A_k f$ have the strong sweeping out property.

Define for each i, $1 \leq i \leq d$,

$$\Omega^i = \{(n_k^i, \ell_k^i): (n_k, \ell_k) \in \Omega\}$$

and

$$\Omega_\alpha^i = \{(z,s): |z-n| < \alpha(s-\ell) \text{ for some } (n,\ell) \in \Omega^i\}.$$

With this notation we can state and prove:

__Theorem 3.2.__ *Assume that for some i, $1 \leq i \leq d$, Ω_α^i does not satisfy a linear growth condition of the form $|\Omega_\alpha^i(\lambda)| \leq c\lambda$ for any choice of $c<\infty$. Then there exists a function in $L^\infty(X)$ such that the averages $A_k f(x)$ diverge for a.e. x. Further, given $\epsilon>0$, there exists a set E of measure less than ϵ, such that $\limsup A_k(\chi_E)(x) = 1$ and $\liminf A_k(\chi_E)(x) = 0$ for a.e. x.*

__Proof.__ We give the details only in the case of two transformations. We can assume with out loss of generality that Ω_α^1 fails to satisfy the linear growth condition. We will use a two dimensional extension of the Kakutani–Rohlin tower construction. Given N, an integer, and small number ϵ there exists a set A such that

a) $\{T_1^i T_2^j A\}$ such that $0 \leq i \leq N$, $0 \leq j \leq N$, are pairwise disjoint sets

and

b) $m(\bigcup_{i=0}^{N} \bigcup_{j=0}^{N} T_1^i T_2^j A) > 1-\epsilon$.

(See [8] or [9] for the details of this construction.)

Take any positive integer p. Because Ω_α^1 fails to satisfy the linear growth condition, we can find a $\lambda = \lambda_p$ such that $p^2 5\lambda_p \leq |\Omega_1^1(\lambda_p)|$. Let ν correspond to the point $(n_\nu^1,\ell_\nu^1) \in \Omega^1$ such that the cone with aperture 1 at (n_ν^1,ℓ_ν^1) contains the point $\sup\{z: z\in\Omega_1^1(\lambda_p)\}$. Select N very large compared both with $5\lambda_p + \sup\{z: z\in\Omega_1(\lambda_p)\}$ and with $n_\nu^2+\ell_\nu^2$.

Using the extension of the Kakutani–Rohlin construction, Form a large Kakutani–Rohlin "rectangle" of sides NxN and error less than $1/N^2$. Define F_p to consist of the rightmost $4\lambda_p+1$ columns of the rectangle. Then $m(F_p) \le (4\lambda_p+1)/N$. Let $x \in X$ be in the "rectangle", exactly $2\lambda_p+1$ steps to the left of the right hand edge, and at bottom edge of the rectangle. Now for $z \in \Omega_1^1(\lambda_p)$ we have $|z-n_r^1| \le (\lambda_p - \ell_r^1)$ for some $(n_r^1,\ell_r^1) \in \Omega^1$, so $\lambda_p \ge \ell_r^1$. Also in this case, $|z-(n_r^1+\ell_r^1)| \le |z-n_r^1| + \ell_r^1 \le \lambda_p + \lambda_p = 2\lambda_p$, so that $|-z+n_r^1| \le \lambda_p$ and $|-z+(n_r^1+\ell_r^1)| \le 2\lambda_p$. This means that

$$T_1^{-z+n_r^1}x \in F_p, \cdots, T_1^{-z+(n_r^1+\ell_r^1)}x \in F_p.$$

Hence for

$$A_r^i f(x) = \frac{1}{\ell^i} \sum_{j=n_r^i}^{n_r^i+\ell_r^i-1} f(T_i^j x)$$

we have

$$A_r^1 \chi_{F_p}(T_1^{-z}x) \ge 1$$

and since this average is exactly the same as $A_r^1 A_r^2 \chi_{F_p}(T_1^{-z}T_2^s x)$ for any s, $0 \le s \le N-(n_\nu^2+\ell_\nu^2)$ we also have

$$\sup_k A_k \chi_{F_p}(T_1^{-z}T_2^s x) \ge 1$$

for each (z,s) such that $z \in \Omega_1^1(\lambda_p)$ and $0 \le s \le N-(n_\nu^2+\ell_\nu^2)$. Let

$$E_p = \{x: \sup_k A_k \chi_{F_p}(x) \ge 1 - \epsilon\}.$$

We have

$$(N-(n_\nu^2+\ell_\nu^2))\times|\Omega_1^1(\lambda_p)|/(2N^2) \le m(\sup_k A_k\chi_{F_p}(x) \ge 1-\epsilon) = m(E_p)$$

but

$$(N-(n_\nu^2+\ell_\nu^2))\times|\Omega_1^1(\lambda_p)|/(2N^2) \ge (N-(n_\nu^2+\ell_\nu^2))p^2 5\lambda_p/(2N^2)$$

$$\ge p^2/2(\frac{N-(n_\nu^2+\ell_\nu^2)}{N})m(F_p)$$

$$\ge m(F_p)p^2/4.$$

Thus $m(E_p)/m(F_p) \ge p^2/4$. Note that we could also apply the above argument to the set Ω with the first few terms removed. Thus the conditions of Theorem 3.1 are satisfied. Applying this theorem completes the proof. □

<u>Example:</u> Consider averages of the form

$$\frac{1}{N}\sum_{k=N^2}^{N^2+N} \frac{1}{2^{2^N}}\sum_{j=0}^{2^{2^N}-1} f(T_1^kT_2^jx),$$

where T_1 and T_2 are induced by ergodic measure preserving point transformations. Clearly for most values of x, and N large, the inner sum should be close to the integral of f. On the other hand, if we were to remove the inner average, we could find functions such that the outer average diverges for almost every x. The question is "does the inner average improve the function enough to make the outer average converge?" Theorem 3.2 shows that any average of this form must diverge, and in fact have the "strong sweeping out property". More generally we have for any increasing sequence B(k), that the averages

$$A_kf(x) = \frac{1}{\ell_k}\sum_{i=n_k}^{n_k+\ell_k} \frac{1}{B(k)}\sum_{j=0}^{B(k)-1} f(T^iS^jx).$$

have the strong sweeping out property whenever the sequence $\{(n_k,\ell_k)\}_{k=1}^{\infty}$ fails to have the linear growth condition. This example shows how badly the usual Cesaro averages can behave. There is a set A with $m(A)<\epsilon$ such that For $f = \chi_A$ with $m(A) < \epsilon$, the inner average must be close to 1 on a large enough set so that when we compute the outer average, the double average can be close to 1 infinitely often. This happens even though for a.e. x, we know that the inner average converges to $m(A) < \epsilon$.

References

1. Akcoglu, M., and del Junco, A., *Convergence of averages of point transformations*, **Proc. A.M.S.**, **49**(1975) 265–266.

2. Asmar, N., Berkson, E., and Gillespie, T., *Transfert des inégalités maximales de type faible*, **C. R. Acad. Sci. Paris**, **310**(1990) 167–170.

3. Bellow, A., Jones, R. and Rosenblatt, J., *Convergence for moving averages*, **Ergodic Theory and Dynamical Systems**, to appear.

4. Bellow, A., Jones, R., and Rosenblatt, J., *Harmonic Analysis and ergodic theory", in Almost Everywhere Convergence*, in Almost Everywhere Convergence, Proceedings of the International Conference on Almost Everywhere Convergence in Probability and Ergodic Theory, Columbus Ohio, G. Edgar and L. Sucheston editor, Academic Press, 1989.

5. Calderón, A.P., *Ergodic theory and translation–invariant operators*, **Proc. Nat. Acad. Sci.**, **59**(1968) 349–353.

6. Frangos, N. E., and Sucheston, L., *On multi parameter ergodic and martingale theorems in infinite measure spaces*, **Probability Theory and Related Fields**, **71**(1986) 477–490.

7. Jones, R., and Olsen, J., *Subsequence ergodic theorems for operators*, pre–print.

8. Katznelson, I., and Weiss, B., *Commuting measure preserving transformations*, **Israel J. of Math.**, **12**(1972) 161–173.

9. Lind, D., *Locally compact measure preserving flows*, **Advances in Math.**, 15(1975) 175–193.

10. Nagel, A., and Stein, E. M., *On certain maximal functions and approach regions*, **Advances in Mathematics**, **54**(1984), 83–106.

11. Olsen, J., *Dominated estimates of convex combinations of commuting isometries*, **Israel J. Math.**, **11**(1972) 1–13.

12. Olsen, J., *Multi–Parameter weighted ergodic theorems from their single parameter versions*, in Almost Everywhere Convergence, Proceedings of the International Conference on Almost Everywhere Convergence in Probability and Ergodic Theory, Columbus Ohio, G. Edgar and L. Sucheston editor, Academic Press, 1989.

13. Stein, E. M., Singular Integrals and Differentiability properties of Functions, Princeton University Press, Princeton, N.J. 1970.

14. Sucheston, L., *On one–parameter proof of almost sure convergence of multi parameter processes*, **ZW.**, **63**(1983), 43–49.

15. Sueiro, J., *A note on maximal operators of Hardy–Littlewood type*, **Math. Proc. Camb. Phil. Soc.**, **102**(1987), 1213–1216.

16. Zygmund, A., *An individual ergodic theorem for non–commutative transformations*, **Acta. Sci. Math. (Szeged)** **14**(1951) 103–110.

An Almost Sure Convergence Theorem For Sequences of Random Variables Selected From Log-Convex Sets

by John C. Kieffer*

Electrical Engineering Department
University of Minnesota, Twin Cities
200 Union Street SE
Minneapolis, Minnesota 55455

* Supported by NSF Grant NCR-8702176.

ISBN 0-12-085520-8

1. Log-Convex Sets of Random Variables

Let F be a nonempty set of random variables which are defined on the same probability space. Suppose that each random variable in F has finite expectation and that $B(F) > -\infty$, where, throughout this paper, B(F) is the number defined by

$$B(F) = \inf\{EY: Y \in F\}$$

Such a set F can arise naturally, for example, in the context of a type of constrained extremum problem of information theory (Kieffer, 1991). In such a context one may wish to know whether F satisfies the property that there exists a unique random variable $Y^* = Y^*(F)$ in the L^1 closure of F such that $EY^* = B(F)$. Theorem 1, stated and proved at the end of this section, gives a sufficient condition on F for this property to hold. This condition is that F be log-convex. As defined in (Kieffer, 1989), F is <u>log-convex</u> if the set $\{e^{-Y}: Y \in F\}$ is convex.

Given a sequence of log-convex sets $\{F_n\}$, Theorem 1 gives us the existence of the sequence of random variables $\{Y^*(F_n)\}$. It is then a natural question to ask whether this sequence, suitably normalized, converges almost surely. The main result of this paper, Theorem 2, will give us such a convergence theorem. Theorem 2 is stated and proved in Section 2. We conclude the paper in Section 3 with two examples of sequences of log-convex sets of random variables; these examples yield, via Theorem 2, the subadditive ergodic theorem of Kingman (1968) on the one hand and a generalized Shannon-McMillan-Breiman theorem on the other hand.

<u>Theorem 1</u>. Let F be a nonempty set of random variables defined on the same probability space, each of which has finite expectation. Assume that F is log-convex and that $B(F) > -\infty$. Then, up to almost sure equivalence, there is a unique

Y^* in the L^1 closure of F such that $EY^* = B(F)$. Furthermore, Y^* is also the unique element in the L^1 closure of F satisfying

$$E[e^{Y^*-Y}] \leq 1, \; Y \in F \tag{1}$$

Finally, the following two inequalities are valid:

(i) $\Pr[Y \geq Y^* - \varepsilon] \geq 1 - e^{-\varepsilon}, \; Y \in F, \; \varepsilon > 0.$

(ii) $E[|Y - Y^*|] \leq 2[EY - EY^*] + \sqrt{2}\,\sqrt{EY - EY^*}, Y \in F.$

We prove Theorem 1 by means of two lemmas.

Lemma 1. Let U be a random variable such that $E[|U|] < \infty$ and $E[e^{-U}] \leq 1$. Then $E[U] \geq 0$ and

$$E[|U|] \leq 2E[U] + \sqrt{2}\,\sqrt{E[U]}. \tag{2}$$

Proof. One concludes $E[U] \geq 0$ using Jensen's inequality. Let $(\Omega, \mathfrak{F}, P)$ be the probability space on which U is defined. Let $\beta = E[e^{-U}]$, and let Q be the probability measure on $\mathfrak{F}$ such that

$$Q(E) = \beta^{-1} \int_E e^{-U}\, dP, \; E \in \mathfrak{F}.$$

Barron (1986) showed that

$$\int \left|\log \frac{d\lambda_1}{d\lambda_2}\right| d\lambda_1 \leq \int \left(\log \frac{d\lambda_1}{d\lambda_2}\right) d\lambda_1 + \sqrt{2}\,\sqrt{\int \left(\log \frac{d\lambda_1}{d\lambda_2}\right) d\lambda_1}, \tag{3}$$

for any pair of probability measures λ_1, λ_2 such that λ_1 is absolutely continuous with respect to λ_2. (The logarithms in Inequality (3), and elsewhere in the paper, are natural.) We now show that Inequality (2) follows from Inequality (3), if we take $\lambda_1 = P$, $\lambda_2 = Q$. Since $dP/dQ = \beta e^U$,

$$E[U] = \int (\log \frac{dP}{dQ})\, dP + \log(\beta^{-1}). \tag{4}$$

Both terms on the right side of (4) are non-negative, so

$$\int (\log \frac{dP}{dQ})\, dP \leq E[U]\,, \tag{5}$$

$$\log(\beta^{-1}) \leq E[U]\,. \tag{6}$$

Also,

$$|U| \leq \log(\beta^{-1}) + |\log \frac{dP}{dQ}|\,. \tag{7}$$

Taking the expected value of both sides of Inequality (7) and then applying Relations (6), (3), and (5), one obtains Inequality (2).

Our second lemma is due to Bell and Cover (1988).

Lemma 2. Let F be a log-convex set of random variables having finite expectation. Then, the following two statements are equivalent for a random variable $Y^* \in F$:

(i) $E[e^{Y^*-Y}] \leq 1$, $Y \in F$, and

(ii) $E[Y^*] = B(F)$.

Proof of Theorem 1. We first show the existence of Y^* in the L^1 closure of F such that $EY^* = B(F)$. Pick a sequence $\{Y_n: n=1,2,...\}$ of random variables from F such that $EY_n \to B(F)$. For each n, let F_n be the log-convex set of all random variables of the form

$$-\log\left(\sum_{i=1}^{n} \alpha_i e^{-Y_i}\right),$$

where $\alpha_1,...,\alpha_n$ are non-negative numbers summing to one. Note that $F_n \subset F$, $n \geq 1$. Using the dominated convergence theorem, it is easy to show that there exists for each n a $Y_n^* \in F_n$ such that $EY_n^* = B(F_n)$. Since $B(F_n) \to B(F)$, existence of Y^* will follow if we can show L^1 convergence of $\{Y_n^*\}$. Applying Lemma 2, and the fact that $Y_m^* \in F_n^*$ for $m < n$, we obtain

$$E[e^{Y_n^* - Y_m^*}] \leq 1, n > m.$$

Applying Lemma 1,

$$E[|Y_n^* - Y_m^*|] \leq 2[EY_n^* - EY_m^*] + \sqrt{2}\sqrt{EY_n^* - EY_m^*}, n > m.$$

The preceding inequality gives us L^1 convergence of $\{Y_n^*\}$, via the Cauchy criterion.

To obtain uniqueness of Y^*, let $\overline{F}$ denote the L^1 closure of F, and let Y_1, Y_2 be two random variables in $\overline{F}$ whose expected value is B(F). Let Y be the random variable

$$Y = -\log\left[\frac{1}{2}e^{-Y_1} + \frac{1}{2}e^{-Y_2}\right]$$

It can be shown that $Y \in \overline{F}$ and therefore

$$EY \geq B(F) \tag{8}$$

The strict concavity of the logarithm function gives us

$$EY < \frac{1}{2}(EY_1 + EY_2) = B(F) \tag{9}$$

if $\Pr[Y_1 \neq Y_2] > 0$. Since Inequalities (8) and (9) are incompatible, we must have $Y_1 = Y_2$ a.s. .

Let Y^* be the unique element of $\overline{F}$ satisfying $EY^* = B(F) = B(\overline{F})$. Since $\overline{F}$ is log-convex, we may apply Lemma 2 to conclude that Inequality (1) holds. Conversely, let Y^* be an element of $\overline{F}$ for which Eq. (1) holds. Then, by Lemma 1, $EY \geq EY^*$ for every $Y \in F$, whence Y^* must be the unique element of $\overline{F}$ satisfying $EY^* = B(F)$. Thus, the unique $Y^* \in \overline{F}$ satisfying $EY^* = B(F)$ coincides with the unique $Y^* \in \overline{F}$ satisfying Inequality (1).

Finally, to complete the proof of Theorem 1, we point out that Statement (i) follows from Inequality (1) and that Statement (ii) follows from Lemma 1.

Notation. From now on, if F is a set of random variables satisfying the hypotheses of Theorem 1, we will let $Y^*(F)$ denote the unique random variable in the L^1 closure of F whose expected value is B(F).

2. Derivation of Main Result

We fix throughout this section a probability space $(\Omega,\mathcal{F},P)$ and a measurable, measure-preserving transformation $T: \Omega\rightarrow\Omega$. Let $M(\Omega)$ denote the set of all extended real-valued random variables defined on Ω, and let $L^1(\Omega)$ denote the set consisting of all real-valued random variables in $M(\Omega)$ having finite expectation. The transformation T induces an operator on $M(\Omega)$ (which we shall also denote by T) as follows:

$$(TY)(\omega) = Y(T\omega),\ \omega\in\Omega,\ Y\in M(\Omega).$$

Definition. Let $\{F_n\colon\ n = 1,2,\ldots\}$ be a sequence of nonempty subsets of $L^1(\Omega)$. We say that the sequence $\{F_n\}$ is T-additive if

$$\{Y_n + T^nY_m\colon Y_n\in F_n, Y_m\in F_m\}\subset F_{n+m},\ n,m = 1,2,\ldots .$$

Here is our main result.

Theorem 2. For $n = 1,2,\ldots$, let F_n be a nonempty, log-convex subset of $L^1(\Omega)$ such that $B(F_n) > -\infty$. Suppose the sequence $\{F_n\}$ is T-additive. Let $B = \inf_n n^{-1} B(F_n)$. Then, $\{Y^*(F_n)/n\}$ converges almost surely to a T-invariant $Y^*\in M(\Omega)$ whose expected value is B. If $B > -\infty$, then $Y^*\in L^1(\Omega)$ and $\{Y^*(F_n)/n\}$ also converges in L^1 mean to Y^*.

We shall prove Theorem 2 via a sequence of five lemmas. In the subsequent development, we use Y_n^* as an abbreviation for $Y^*(F_n)$, $n \geq 1$. The mean convergence of $\{Y_n^*/n\}$ is much easier to prove than almost sure convergence; consequently, we have put the lemmas in an order so that the mean convergence can be deduced after just the first two lemmas.

Lemma 3. Let $B > -\infty$. If $Y_n \in F_n$ $(n \geq 1)$, then

$$\frac{1}{2}\, \overline{\lim}\, E\left[\left| \frac{Y_n}{n} - \frac{Y_n^*}{n} \right|\right] \leq \{\overline{\lim}\, E[\frac{Y_n}{n}]\} - B.$$

Proof. Result follows from Statement (ii) of Theorem 1.

Lemma 4. For each positive integer k, there exists $Y_n \in F_n$ $(n \geq 1)$ such that the sequence $\{Y_n/n\}$ converges almost surely and in L^1 mean to a random variable with expected value $B(F_k)/k$.

Proof. Result follows from the Birkhoff-Von Neumann ergodic theorem, coupled with the fact that $\{F_n\}$ is T-additive.

Remark. We can see now that $\{Y_n^*/n\}$ converges in L^1 mean, provided $B > -\infty$. The reason for this is that, if we couple Lemmas 3 and 4, we see that $\{Y_n^*/n\}$ is a Cauchy sequence in L^1 mean.

Lemma 5. If $Y_n \in F_n$ $(n \geq 1)$, then

$$\overline{\lim}\, n^{-1} Y_n^* \leq \overline{\lim}\, n^{-1} Y_n \quad \text{a.s., and}$$

$$\underline{\lim}\, n^{-1} Y_n^* \leq \underline{\lim}\, n^{-1} Y_n \quad \text{a.s.}.$$

Proof. Result follows from Statement (i) of Theorem 1.

Lemma 6. $E[\overline{\lim}\, n^{-1}Y_n^*] \le B$.

Proof. Result follows from Lemmas 3 and 4.

Lemma 7. The random variable $\underline{\lim}\, n^{-1}Y_n^*$ is T-invariant.

Proof. Applying Lemma 5 to the sequence $Y_n = Y_1^* + TY_{n-1}^*$ $(n \ge 1)$, we obtain

$$\underline{\lim}\, n^{-1} Y_n^* \le T(\underline{\lim}\, n^{-1} Y_n^*) \quad \text{a.s.},$$

from which the result follows.

The last lemma we present (Lemma 8) is the crucial lemma for proving almost sure convergence of the sequence $\{Y_n^*/n\}$. We need to introduce some definitions and notation before presenting Lemma 8.

A partition of a positive integer n is a sequence $(k_1,k_2,\ldots,k_t)$ of positive integers whose sum is n.

For each partition $\pi = (k_1,\ldots,k_t)$ of the positive integer n, let Y_π^* denote the random variable in F_n given by

$$Y_\pi^* = Y_{k_1}^* + T^{k_1} Y_{k_2}^* + T^{k_1+k_2} Y_{k_3}^* + \ldots + T^{k_1+\ldots+k_{t-1}} Y_{k_t}^*.$$

If N,n are positive integers, let S(n,N) be the set of all partitions π of n such that each entry of π is either 1 or is at least N. Let $|S(n,N)|$ be the number of partitions in S(n,N), and let α_N be the number

$$\alpha_N = \overline{\lim_{n\to\infty}}\; n^{-1} \log|S(n,N)|, \quad N = 1,2,\ldots.$$

We remark for later use that $\alpha_N \to 0$ as $N \to \infty$.

For positive integers n,N, define Z(n,N) to be the element of F_n given by

$$Z(n,N) = -\log\Big[|S(n,N)|^{-1} \sum_{\pi \in S(n,N)} e^{-Y^*_\pi}\Big] . \tag{10}$$

Lemma 8. For each N = 1,2,...,

$$\overline{\lim_{n\to\infty}}\ n^{-1} Z(n,N) \leq \alpha_N + \underline{\lim}_{n\to\infty}\ n^{-1} Y^*_n \quad \text{a.s.} . \tag{11}$$

Before proceeding with the proof of Lemma 8, we show how it, in conjunction with the previous lemmas, can be used to complete the proof of Theorem 2. First, we apply Lemma 5 together with the fact that $\alpha_N \to 0$ to the conclusion of Lemma 8 to obtain

$$\overline{\lim}\ n^{-1} Y^*_n \leq \underline{\lim}\, n^{-1} Y^*_n \quad \text{a.s.} .$$

This gives us almost sure convergence of $\{Y^*_n/n\}$. From Lemma 7, the limit function is T-invariant. If $B = -\infty$, we see from Lemma 6 that the expected value of the limit function must be B. If $B > -\infty$, we know already that $\{Y^*_n/n\}$ converges in L^1 mean, so that the expected value of the limit function must be B in this case also. The proof of Theorem 2 is thus complete once we prove Lemma 8.

Proof of Lemma 8. The method of proof is inspired by the method developed by Ornstein and Weiss (1983). We fix a positive integer N throughout the proof and demonstrate (11) for this N. Let W be an arbitrary T-invariant random variable from $L^1(\Omega)$ such that

$$P[\underline{\lim}\, n^{-1} Y_n^* < W] = 1. \tag{12}$$

(There exists at least one such random variable W, namely, the limit of the sequence $\{n^{-1} \sum_{p=o}^{n-1} T^p Y_1^*\}$, increased by one.)

Let J be a positive integer such that $J \geq N$, and let E_J be the event

$$E_J = \bigcup_{i=N}^{J} \{Y_i^* < iW\}. \tag{13}$$

For each p = 0,1,..., let $E_J(p)$ be the event

$$E_J(p) = \bigcup_{i=N}^{J} \{T^p Y_i^* < iW\}. \tag{14}$$

Let ω be a fixed, but arbitrary, element of Ω. Let $k_1, k_2, \ldots$ be the sequence of positive integers defined inductively as follows:

(i) If $\omega \in E_J$, then k_1 is the smallest $k (N \leq k \leq J)$ such that $Y_k^*(\omega) < kW(\omega)$; if $\omega \notin E_J$, $k_1 = 1$.

(ii) To define k_i for $i > 1$ assuming that $k_1, \ldots, k_{i-1}$ have been defined, let $p_i = k_1 + \ldots + k_{i-1}$. If $\omega \in E_J(p_i)$, k_i is the smallest $k (N \leq k \leq J)$ such that $T^k Y_k^*(\omega) < kW(\omega)$; if $\omega \notin E_J(p_i)$, $k_i = 1$.

Let n be a positive integer. We define a partition $\pi_n \in S(n,N)$. Find the largest t such that $k_1 + ... + k_t \leq n$. Then .

$$\pi_n = (k_1,...,k_t), \text{ if } k_1 + ... k_t = n$$
$$= (k_1,...,k_t,1,1,...1), \text{ otherwise.}$$

It is easy to see from Eq. (10) that

$$Z(n,N)(\omega) \leq \log|S(n,N)| + Y^*_{\pi_n}(\omega) . \tag{15}$$

Also, one can see that

$$Y^*_{\pi_n}(\omega) \leq nW(\omega) + \sum_{p=o}^{n-1} I[E^c_J(p)](\omega)\{|(T^pY^*_1)(\omega)| + |W(\omega)|\}$$

$$+ \sum_{p=n-J}^{n-1} \{|(T^pY^*_1)(\omega)| + |W(\omega)|\}, \tag{16}$$

where $E^c_J(p)$ is the complement of the event $E_J(p)$ and $I[E^c_J(p)]$ is the indicator function of $E^c_J(p)$. Comparing Eq. (13) and Eq. (14) and using the fact that W is T-invariant, we have

$$T^p I[E^c_J] = I[E^c_J(p)] \quad \text{a.s., } p = 0,1,... .$$

This relation allows us to deduce the behavior of the right side of Inequality (15) as $n \to \infty$ for almost every $\omega \in \Omega$, via the pointwise ergodic theorem. Letting $\mathcal{I}$ be the sigma-field generated by the T-invariant random variables in $M(\Omega)$, we conclude from Eq. (14) and Inequality (15) that

$$\overline{\lim_{n\to\infty}}\ n^{-1} Z(n,N) \le \alpha_N + W + |W| P(E_J^c|\mathcal{I}) +$$

$$E(Y_1^* I[E_J^c]|\mathcal{I}) \quad \text{a.s.,} \quad J = 1,2,\ldots . \tag{17}$$

In view of Eq. (11) and Eq. (12), $P(E_J^c) \to 0$ as $J \to \infty$. This fact, applied to Inequality (17), tells us that

$$\overline{\lim_{n\to\infty}}\ n^{-1} Z(n,N) \le \alpha_N + W \quad \text{a.s.,} \tag{18}$$

for every T-invariant $W \in L^1(\Omega)$ satisfying Eq. (12). We may choose a strictly decreasing sequence $\{W_k\}$ of T-invariant random variables from $L^1(\Omega)$ which converges to $\underline{\lim}\ n^{-1} Y_n^*$ a.s.. (Since $\underline{\lim}\ n^{-1} Y_n^*$ is bounded above by a T-invariant random variable in $L^1(\Omega)$, and is itself T-invariant, there must exist such a sequence $\{W_k\}$.) Since Inequality (18) is valid with $W = W_k$ for every k, we conclude from Inequality (17) that Inequality (11) is true, thereby completing the proof of Lemma 8.

3. Examples

Example 1. Let $(A,\mathcal{A})$ be a measurable space and let $X_1,X_2,\ldots$ be a stationary random sequence with state space $(A,\mathcal{A})$. Let $Y_1,Y_2,\ldots$ be a sequence of L^1 random variables in which each Y_n is of the form

$$Y_n = f_n(X_1,X_2,\ldots)$$

where f_n is a measurable function on the unilateral sequence space formed from $(A,\mathcal{A})$. Suppose that

$$Y_n + f_m(X_{n+1},X_{n+2},\ldots) \geq Y_{n+m} \text{ a.s., } n,m = 1,2,\ldots$$

The subadditive ergodic theorem of Kingman (1968) then tells us that $\{Y_n/n\}$ converges almost surely. We can also deduce this as a special case of Theorem 2 by taking each F_n in Theorem 2 to be the log-convex set of all L^1 random variables Y based on $X_1,X_2,\ldots$ such that $Y \geq Y_n$ a.s.. (One sees that $Y^*(F_n) = Y_n$ for every n.)

Example 2. Again, let $(A,\mathcal{A})$ be a measurable space and let $X_1,X_2,\ldots$ be a stationary random sequence with state space $(A,\mathcal{A})$. Let π: $A \times \mathcal{A} \rightarrow [0,1]$ be a regular conditional probability. For each n = 1,2,..., let $\mathcal{A}^n$ be the n-fold product sigma-field formed from $\mathcal{A}$, and let π^n: $A \times \mathcal{A}^n \rightarrow [0,1]$ be the regular conditional probability given by

$$\pi^n(x,d(x_1,\ldots,x_n)) = \pi(x_{n-1},dx_n)\ldots\pi(x_1,dx_2)\pi(x,dx_1).$$

Let Q_1 be a probability measure on $\mathcal{A}$ and for n = 2,3,... let Q_n be the probability measure on $\mathcal{A}^n$ such that

$$Q_n(d(x_1,...,x_n)) = \pi^n(x_1,d(x_2,...,x_n))Q(dx_1).$$

Let P_n be the probability distribution of $(X_1,...,X_n)$ on $\mathcal{A}^n$ and suppose P_n is absolutely continuous with respect to Q_n, $n \geq 1$. For each n, let g_n: $A^n \rightarrow [0,\infty)$ be the Radon-Nikodym derivative of P_n with respect to Q_n. Suppose that $\sup_n\{n^{-1} E[\log g_n(X_1,...,X_n)]\} < \infty$. Then, a recent generalization of the Shannon-McMillan theorem due to Barron (1985) and Orey (1985) states that $\{n^{-1} \log g_n(X_1,...,X_n)\}$ converges almost surely and in L^1 mean to a T-invariant random variable. We can easily deduce this result by applying Theorem 2 to the right sequence $\{F_1,F_2,...\}$. Define F_1 to be the log-convex set of all L^1 random variables Y_1 which are non-negative functions of X_1. For $n \geq 2$, define F_n to be the log-convex set of all L^1 random variables $f_n(X_1,X_2,...,X_n)$ in which f_n: $A^n \rightarrow \mathbf{R}$ is $\mathcal{A}^n$-measurable and satisfies

$$\int_{A^{n-1}} e^{-f_n(x_1,...,x_n)} \pi^{n-1}(x_1,d(x_2,...,x_n)) \leq 1, \quad x_1 \in A.$$

For $n \geq 1$, $Y^*(F_n) = \log g_1(X_1) - \log g_n(X_1,...,X_n)$ a.s. . From Theorem 2, we obtain convergence of $\{Y^*(F_n)/n\}$, which yields convergence of $\{n^{-1} \log g_n(X_1,...,X_n)\}$.

The paper (Kieffer, 1991) gives further examples of almost sure convergence theorems obtainable via choice of $\{F_n\}$ in Theorem 2.

Acknowledgement. The author is grateful to Paul Algoet for bringing the paper (Bell and Cover, 1988) to his attention, and to Andrew Barron and Jingou Liu for helpful comments on an earlier version of this paper.

References

Barron, A. (1985). The strong ergodic theorem for densities: Generalized Shannon-McMillan-Breiman theorem. Ann. Probab. 13 1292-1303.

Barron, A. (1986). Entropy and the Central Limit Theorem. Ann. Probab. 14 336-342.

Bell, R. and Cover, T. (1988). Game-theoretic optimal portfolios. Management Science 34 724-733.

Kieffer, J. (1989). An ergodic theorem for constrained sequences of functions. Bull. Amer. Math. Soc. 21, 249-254.

Kieffer, J. (1991). Sample converses in rate distortion theory. To appear, IEEE Trans. Inform. Theory.

Kingman, J. (1968). The ergodic theory of subadditive stochastic processes. J. Roy. Stat. Soc. B 30 499-510.

Orey, S. (1985). On the Shannon-Perez-Moy theorem. Contemp. Math. 41 319-327.

Ornstein, D. and Weiss, B. (1983). The Shannon-McMillan-Breiman Theorem for a class of amenable groups. Israel J. Math. 44 53-60.

DIVERGENCE OF ERGODIC AVERAGES AND ORBITAL CLASSIFICATION OF NON-SINGULAR TRANSFORMATIONS

I. Kornfeld

North Dakota State University

The problem considered in this paper is connected with the relationship between the convergence (or divergence) properties of the averages for non-singular transformations of a measure space and the classification of such transformations up to orbital equivalence [5].

We start with Chacon's example of divergence of ergodic averages ([1], cf also [4], p. 151). Let $(\Omega,\mathcal{B},\mu)$ be a Lebesgue space with non-atomic probability measure (which is isomorphic to [0,1] with Lebesgue measure). A non-singular invertible ergodic transformation $\tau\colon \Omega\ \Omega$ is constructed in [1] such that for some $f\in L^1_+(\mu)$ the averages

$$A_n(f) = A_n(f;\tau) = \frac{1}{n}\sum_{k=0}^{n-1} T^k f$$

diverge μ-a.e. on Ω. Here $T = T(\tau)$ is the positive L^1-contraction induced by τ and defined by

$$(Tf)(x) = \frac{d\tau^{-1}\mu}{d\mu}(x)\cdot f(\tau^{-1}x),\quad x\in\Omega.$$

In fact, the function f even satisfies

$$\liminf A_n(f)(x) = 0 \qquad \text{a.e.}$$

$$\limsup \frac{1}{n}(T^n f)(x) = \infty \qquad \text{a.e.}$$

It follows from the Chacon-Ornstein theorem that the divergence of $A_n(f)$ in Chacon's example holds a.e. for all $f\in L'_+$, $f\neq 0$. It is also clear that the transformation τ cannot admit a σ-finite invariant measure $\mu_0 \sim \mu$, in other words, τ is of type III in Krieger-Araki-Woods classification [5].

ISBN 0-12-085520-8

It seems natural to ask whether or not such divergence holds for all type III transformations. This question was answered negatively by an example of H. Weiner [8]. Another, more general construction was given by Y. Ito [9].

We will be interested in the question that can be informally stated as follows: are there any orbital restrictions for such "divergent" behavior? More precisely: given a non-singular invertible ergodic transformation τ, is it possible to find $\sigma \sim \tau$ (σ orbitally equivalent to τ) such that the averages $A_n(f,\sigma)$ diverge μ-a.e. for some $f \in L^1_+(\mu)$? Another and, maybe, more interesting open question is: given τ, is it possible to find $\sigma \sim \tau$ for which $A_n(f;\sigma)$ converge a.e. for all $f \in L^1$?

Definitions and conventions. We assume all subsets introduced below to be measurable and all expressions about subsets to hold μ-mod 0. Let G be the group of all non-singular invertible bimeasurable transformations of $(\Omega, \mathcal{B}, \mu)$. For $\tau \in G$ and $x \in \Omega$ let $\mathrm{Orb}_\tau(x) = \{\tau^n x: n \in \mathbb{Z}\}$. For $\tau \in G$ the full group $[\tau]$ is defined by

$$[\tau] = \{\sigma \in G: \mathrm{Orb}_\sigma(x) \subseteq \mathrm{Orb}_\tau(x) \text{ for all } x \in \Omega\}.$$

Two transformations τ_1, τ_2 defined on $(\Omega_1, \mathcal{B}_1, \mu_1)$, $(\Omega_2, \mathcal{B}_2, \mu_2)$ respectively are called weakly (or orbitally) equivalent (notation: $\tau_1 \overset{w}{\sim} \tau_2$) if there exists an isomorphism η: $(\Omega_1, \mathcal{B}_1, \mu_1)$ $(\Omega_2, \mathcal{B}_2, \mu_2)$ such that $\eta \, \mathrm{Orb}_{\tau_1}(x_1) = \mathrm{Orb}_{\tau_2}(\eta x_1)$ for all $x_1 \in \Omega_1$.

Two subsets $A, B \in \mathcal{B}$ are called equivalent under the full group $[\tau]$ of a transformation τ (notation: $A \overset{[\tau]}{\sim} B$) if there exists $\sigma \in [\tau]$ for which $B = \sigma A$.

We recall the definition of the asymptotic ratio set $r(\tau)$ for a non-singular ergodic transformation τ (cf[5]):

$$r(\tau) = \Big\{ a\in\mathbb{R}^1_+ : \forall\varepsilon>0,\ \forall A\in\mathcal{B}, \mu(A)>0,$$

$$\exists B\subset A,\ \mu(B)>0, \text{ and } \exists i\in\mathbb{Z}:$$

$$\tau^i B\subseteq A \text{ and } \left|\frac{d\tau^i\mu}{d\mu}(x) - a\right|<\varepsilon \text{ for all } x\in B\Big\}$$

A transformation τ admits a σ-finite invariant measure $\mu_0 \sim \mu$ iff $r(\tau) = \{1\}$. In this case τ belongs to type II_1 of Krieger's classification if μ_0 is finite, and to type II_∞ if μ_0 is infinite. The other possibilities are:

type III_λ, $0<\lambda<1$: $r(\tau) = \{\lambda^n : n\in\mathbb{Z}\}\cup\{0\}$,

type III_1: $r(\tau) = [0,\infty]$; type III_0: $r(\tau) = \{0; 1\}$.

THEOREM. For any transformation τ of type III_λ, $\lambda\neq 0$, there exists σ, $\sigma\in[\tau]$, $\sigma \overset{w}{\sim} \tau$, such that for some $f\in L^1_+(\Omega,\mathcal{B},\mu)$ the averages $A_n(S; f)$ diverge μ-a.e., where $S = S(\sigma)$ is the L^1-contraction induced by σ.

<u>Remark</u>. In the case of III_1 - transformations the statement of the theorem can be obtained as a corollary of the following known results:
1) for any type III_λ, $\lambda\neq 0$, the asymptotic ratio set is a complete invariant of orbital equivalence [6];

2) the set of non-singular transformations for which the divergence of averages holds is a residual set in G equipped with the coarse topology [3];

3) the set of type III_1 transformations is also a residual set in G [2].

We use an explicit construction below not only because the case $0<\lambda<1$ is not that straightforward but (mainly) because there is a hope that the restriction $\lambda\neq 0$ is not necessary for the above theorem, and in this general setting an "explicit" approach seems to be natural. Also, the method of the proof shows that many "cutting and stacking constructions" (not only Chacon's example) can be "embedded" into a given orbital class.

For the details of Chacon's example we refer to [4], and we will describe mainly the additional steps and necessary modifications.

LEMMA 1 ([5], cf also [7]). If τ is an ergodic type III transformation, then any two sets A, B of positive measure are equivalent w.r.t. $[\tau]$.

We sketch a simple proof of this known result. We will need later a modification of this lemma (Lemma 4) which can be proved using the same ideas.

<u>Proof of Lemma 1</u>. Let us call the sets A, B weakly equivalent w.r.t. $[\tau]$ if for some $A'\subset A$, $B'\subset B$ we have $A \overset{[\tau]}{\sim} B'$, $B \overset{[\tau]}{\sim} A'$. Weak equivalence implies equivalence, and the standard proof of the classical Schröder-Bernstein theorem of set theory can be used to show this.

To prove that $A \overset{[\tau]}{\sim} B'$ for some $B'\subset B$ we use ergodicity and choose $A_1\subset A$, $B_1\subset B$ and $n\in\mathbb{Z}$ such that $\tau^n A_1 = B_1$. Without loss of generality we can assume that $\frac{d\tau^n\mu}{d\mu}(x)\le M$ for all $x\in A_1$ and some $M<\infty$ (we can replace A_1 by some subset of A_1, if necessary).

Since τ is of type III, so is the induced transformation

$\tau_{B_1} : B_1 \quad B_1$ (the first return map constructed from τ on B_1). This implies that for any $N>0$ there is a set $B_2\subset B_1$ and $m\in\mathbb{Z}$ such that $\mu(\tau^m_{B_1} B_2)\ge N\cdot\mu(B_2)$: otherwise we would have a finite invariant measure $\mu_0\sim\mu|B_1$ for τ_{B_1}. Since $\tau^m_{B_1}(x) = \tau^{m(x)}x$ for some $m(x)$, $x\in B_1$, we can find $\tilde{B}\subset B_2$ and $s\in\mathbb{Z}$ for which $\tau^s\tilde{B}\subset B_1$ and $\mu\ (\tau^s\tilde{B})\ [\mu(\tilde{B})]^{-1}\ge N$.

Now choose $N = M\cdot\mu(A)\cdot[\mu(B)]^{-1}$, and set $\tilde{A} = \tau^{-n}(\tau^s\tilde{B})$. For $p = n-s$ we have $\tau^p\tilde{A} = \tilde{B}$, $\mu(\tilde{B})\cdot[\mu(\tilde{A})]^{-1}\le\mu(B)\cdot[\mu(A)]^{-1}$. We throw out the sets $\tilde{A}$ from A and $\tilde{B}$ from B, and repeat the construction until we exhaust A and, hence, prove that A is equivalent to a subset of B.

Sketch of the proof of theorem. The proof is inductive and the main tool is the repeated application of the non-singular version of the Rokhlin Lemma (due to Linderholm (cf[3])).

1st step. Apply to τ the Rokhlin-Linderholm Lemma with height 2 and remainder $\frac{1}{2}$. We can obtain the sets E_1, $E_2 \subset \Omega$ such that $E_1 \cap E_2 = \emptyset$, $\tau E_1 = E_2$, $\frac{1}{2} \le \mu(E_1 \cup E_2) \le \frac{3}{4}$.

Let $\Omega_1 = E_1 \cup E_2$, and for any $x \in E_1$ define σx to be τx.

Define $f(x) = \chi_{E_1}(x)$, $x \in \Omega$.

n-th step (inductive assumptions). Suppose that after the n-th step we have constructed a set $\Omega_n \subset \Omega$ such that:

i_n) $1 - \frac{1}{2^n} \le \mu(\Omega_n) \le 1 - \frac{1}{2^{n+1}}$;

ii_n) $\Omega_n = E_1^{(n)} \cup \ldots \cup E_{N_n}^{(n)}$ (disj); σ is defined on any $E_i^{(n)}$, $1 \le i < N_n$, and $\sigma(E_i^{(n)}) = E_{i+1}^{(n)}$; (we will call Ω_n the n-th tower);

iii_n) for any $x \in E_i^{(n)}$, $i < N_n$, we have $\sigma x \in \mathrm{Orb}_\tau(x)$;

iv_n) if for $x \in E_1^{(n)}$ we define $\mathrm{Orb}_\sigma^{(n)}(x) = \{\sigma^i x,\ 0 \le i < N_n\}$ (we will call the sets $\mathrm{Orb}_\sigma^{(n)}(x)$ n-fibers), then $\mu(\{x \in \Omega_n : \tau x \notin \Omega_n$, or $\tau x \in \Omega_n$ but x and τx are not in the same n-fiber$\}) \le \frac{\mathrm{const}}{2^n}$, where const does not depend on n;

v_n) $\mu(\{x \in \Omega_n : \inf_k^{(n)} A_k(f;\sigma)(x) \le \frac{1}{n}\}) \ge \mu(\Omega_{n-1}) - \frac{\mathrm{const}}{2^n}$;

vi_n) $\mu(\{x \in \Omega_n : \sup_k \frac{1}{k} S^k f(x) \ge n\}) \ge \mu(\Omega_n) - \frac{\mathrm{const}}{2^n}$;

(in v_n, vi_n the expressions $\inf_k^{(n)}$, $\sup_k^{(n)}$ mean that inf and sup are taken over the values of k for which $A_k(f;\sigma)(x)$, $S^k f(x)$ respectively are defined after the n-th step).

Now let us consider the (n+1)-th step. We split up the rest of the proof into subsections a-e.

a) LEMMA 2. For any $\varepsilon > 0$ there are $M_n = M_n(\varepsilon)$ and a set $\mathcal{N}_n \subset \Omega_n$ consisting of entire n-fibers such that

i) $\mu(\mathcal{N}_n)<\varepsilon$;

ii) for any two points $x,y\in\Omega_n\backslash\mathcal{N}_n$, if x and y are in the same n-fiber and $y = \tau^{k(x,y)}x$, then $|k(x,y)|\le M_n$.

We omit a simple proof of this assertion.

Lemma 2 enables us to find a subset $\Omega'_n\subset\Omega_n$ consisting of the entire n-fibers, $\mu(\Omega_n^1)\ge\mu(\Omega_n) - \frac{1}{2^n}$, and a number $L = L_n$ such that

$$\sup_{x,k} \frac{d\tau^k\mu}{d\mu}(x)\le L \tag{1}$$

where the sup is taken over such pairs (x,k) that $x,\tau^k x$ are in the same fiber of Ω'_n. Now, following the construction of Chacon's example, we add to the n-th tower Ω_n some extra levels $E_i^{(n)}$, $N_n+1\le i<N'_{n+1}$, so as to satisfy i_{n+1}), ii_{n+1}) and iii_{n+1}). Lemma 1 can be used to define the map σ on $\bigcup_{i=N_n}^{N'_n-1} E_i^{(n)}$. Arguing as in ([4]), p. 151), we show that $E_i^{(n)}$, $N_n+1\le i\le N'_{n+1}$, can be chosen such that

$$\inf_k {}^{(n+1)}A_k(f;\sigma)(x)\le \frac{1}{n+1} \tag{2}$$

for all $x\in\Omega'_n$ (The only difference between our construction and the construction in [4] at this point is that in [4] the values $T^i f(x)$, $0\le i\le j$, for $x\in E_j^{(n)}$, $1\le j\le N_n$, are estimated in terms of $\mu(E_j^{(n)})$, $\mu(E_{j-1}^{(n)})$...only, and we have to use (1) in this estimate). The inequality(2) will give us v_{n+1}.

b) Now our concern is to obtain iv_{n+1}. First we apply Lemma 2 to Ω'_{n+1} (instead of Ω_n) with $\varepsilon = \frac{1}{2^{n+1}}$, and obtain a set $\mathcal{N}_{n+1}$, $\mu(\mathcal{N}_{n+1})<\frac{1}{2^{n+1}}$, consisting of the entire fibers, and a constant M satisfying the following condition: for any pair $x,y\in\Omega''_{n+1} = \Omega'_{n+1}\backslash\mathcal{N}_{n+1}$, lying in the same fiber, we have $y = \tau^{k(x,y)}x$ with $|k(x,y)|\le M$. Then we construct a Rokhlin tower for τ with height $M^{(1)} = 5M$ and remainder ε_1 to be specified a bit later. We obtain the representation

$$\Omega = (\bigcup_{j=1}^{M^{(1)}} G_j)\cup R \text{ (disj)}, \text{ where } \mu(R)<\varepsilon_1,\ \tau G_j = G_{j+1},\ 1\le j<M^{(1)}.$$

Let ξ be the partition of Ω with elements G_j, $1\le j\le M^{(1)}$, and R. For any level $E_i = E_i^{(n)}$ of the tower Ω''_{n+1}, consider the partition $\xi_i = \xi|_{E_i}$, as well as the partition η of E_1: $\eta = \bigvee_{i=1}^{N'_{n+1}} \sigma^{-i}\xi_i$. Let $\underline{C}_s$, $1\le s\le S$, be the elements of η. Denote by ζ the partition of Ω''_{n+1} defined by $\zeta|_{E_i} = \sigma^{i-1}\eta$, $1\le i\le N'_{n+1}$.

Remove from E_1 all elements $\underline{C}_s$ of ζ for which $(\bigcup_{i=1}^{N'_{n+1}} \sigma^{i-1}\underline{C}_s)\cap R\ne\emptyset$, and then remove from any E_i, $1\le i\le N^1_{n+1}$, all elements of $\zeta|_{E_i}$ of the form $\sigma^{i-1}\underline{C}_s$, where $\underline{C}_s$ is a removed element from E_1. Retain, for simplicity, the notation Ω''_{n+1} for the remained tower. By choosing ε small enough, we will still have $\mu(\Omega''_{n+1}) \ge\mu(\Omega'_{n+1}) - \frac{1}{2^{n+1}}$. Now Ω''_{n+1} is represented as a disjoint union of "columns" C_s, where $C_s = \bigcup_{i=1}^{N'_{n+1}} \sigma^{i-1}\underline{C}_s$, and $\underline{C}_s$ is a non-removed element of η. For simplicity, we will still denote by S the total number of such elements. The main property of this representation is the following: if x,y are in the same column C_s and $y = \tau^{\pm1}x$, then x,y are in the same fiber. To prove this, we find $x',y',\in\underline{C}_s$ (recall that $\underline{C}_s$ is the bottom level of C_s) such that $x = \tau^{k(x)}x'$. $y = \tau^{k(y)}y'$. This implies $y' = \tau^p x'$, where $|p|\le|k(x)|+|k(y)|+1\le 5M = M^{(1)}$, and $y'\ne x'$ would contradict the construction of the Rokhlin tower.

c) In order to obtain $iv_{n+1})$, we are going to define the maps

$\eta_s : \bar{C}_s \quad \underline{C}_{s+1}$, $1 \le s < S$, from the top of C_s to the bottom of C_{s+1}. These maps will be used as "gluing maps" to construct a single "tall" column out of columns C_s (in other words, we will define $\sigma|_{\bar{C}_s} = \eta_s$, $1 \le s < S$, and, so, use η_s to put C_{s+1} on top of C_s). Let us start with putting C_2 on top of C_1.

LEMMA 3. Let u,v be two points lying in the same fiber of C_2. If $\tau^{\pm 1}u$ and $\tau^{\pm 1}v$ belong to C_1, then they are also in the same fiber of C_1.

Proof. Define $u', v' \in C_1$ by $\tau^{\pm 1}u = \sigma^{i_1}u'$, $\tau^{\pm 1}v = \sigma^{i_2}v'$ for some i_1, i_2. Arguing as before, we will have $v' = \tau^p u'$ with $|p| \le 5M = M^{(1)}$, and $u' \ne v'$ would contradict the construction of the Rokhlin tower.

Now we can define the map η_1 as follows. Let $C_2' \subset C_2$ be the subset of the points $x \in C_2$ for which there exists a point y in the same fiber of C_2 such that τy or $\tau^{-1}y$ belongs to C_1. Consider the point z uniquely defined by $z \in \bar{C}_1$, z and $\tau^{\pm 1}y$ are in the same fiber. We define $\eta_1 z = x$, and Lemma 3 justifies correctness of this definition. η_1 maps a subset $\bar{C}_1'$ of $\bar{C}_1$ to a subset $\underline{C}_2'$ of $\underline{C}_2$. We can usı Lemma 1 and extend η_1 to $(\bar{C}_1 \setminus \bar{C}_1')$, so that to have $\eta_1(\bar{C}_1 \setminus \bar{C}_1') = \underline{C}_2 \setminus \underline{C}_2'$.

Now η_1 is defined completely, and we use it as a "gluing map" to replace the pair of columns C_1, C_2 by a single column D_2 with C_2 on top of C_1. Set, formally, $D_1 = C_1$. Assume, by induction, that we have already constructed the column D_s obtained by putting C_2 on top of C_1, C_3 on top of $C_2, \ldots, C_s$ on top of C_{s-1}. We have to construct now the column D_{s+1} by putting C_{s+1} on top of D_s.

This construction is similar to the construction of the column D_2 out of C_1 and C_2. Namely, we remove from D_s a set $\mathcal{N}^{(s)}$, $\mu(\mathcal{N}^{(s)}) < \frac{1}{2^{n+s}}$, $\mathcal{N}^{(s)}$ consists of the entire fibers, and replace the tower D_s by $D'_s = D_s \backslash \mathcal{N}^{(s)}$ which has the property: if two points $u, v \in C_{s+1}$ are in the same fiber and $\tau^{\pm 1}u$, $\tau^{\pm 1}v \in D'_s$, then $\tau^{\pm 1}u$, $\tau^{\pm 1}v$ are in the same fiber of D'_s.

This property enables us to define correctly the "gluing map" η_s in essentially the same way as η_1 was defined. This completes the s-th step of induction and, thus, proves iv_{n+1}.

d) It remains to prove vi_{n+1}. In order to do this, we will imitate the "cutting part" of the construction of Chacon's example.

Denote by $\bar{\Omega}_{n+1}$ the tower D_s obtained after the last, S-th step of the construction in subsection c). $\bar{\Omega}_{n+1}$ is represented in the form:

$$\bar{\Omega}_{n+1} = G_1 \cup G_2 \cup \ldots \cup G_P \quad \text{(disj)}$$

$$\sigma G_i = G_{i+1}, \quad 1 \le i < P.$$

Here we denote by G_i, $1 \le i \le P$, the levels of the tower D_S.

We repeat the argument from subsection 1 (lemma 3) and throw out from $\bar{\Omega}_{n+1}$ a set $\bar{\mathcal{N}}_{n+1}$ of measure $\le \frac{1}{2^{n+1}}$ consisting of entire fibers and such that there exists $\bar{M} < \infty$ with the property: for any $x, y \in \bar{\Omega}_{n+1} \backslash \bar{\mathcal{N}}_{n+1}$ lying in the same fiber, $y = t^k x$ implies $|k| \le \bar{M}$.

Then, just as in subsection a), we define a number $\tilde{L} = \tilde{L}_{n+1}$ such that on a set $\tilde{\Omega}_{n+1} \subset \bar{\Omega}_{n+1}$, $\mu(\bar{\Omega}_{n+1} \backslash (\tilde{\Omega}_{n+1}) \le \frac{1}{2^{n+1}}$, consisting of entire fibers we will have

$$\sup_{x,k} \frac{d\tau^k \mu}{d\mu}(x) \le \tilde{L}, \tag{3}$$

where sup is taken over the pairs (x,k) such that x and $\tau^k x$ are in the same fiber of $\tilde{\Omega}_{n+1}$.

Now we proceed to the cutting part of the construction. As in [4], p. 152, we define the parameters ε_n, $K = K_n$, $c = c_n$, $\alpha_{n,k}$ $(1\le k\le K)$, and partition the set $\tilde{G}_1 = G_1 \cap \tilde{\Omega}_{n+1}$ (the bottom of the tower $\tilde{\Omega}_{n+1}$) into K subsets

$$\tilde{G}_1 = \tilde{G}_{1,0} \cup \tilde{G}_{1,1} \cup \ldots \cup \tilde{G}_{1,K-1} \quad \text{(disj)}$$

such that $\mu(\tilde{G}_{1,k}) = \alpha_{n,k} \cdot \mu(\tilde{G}_1)$. The choice of the parameters differs only slightly from the choice in [4]. Namely, we set $\varepsilon_n = \dfrac{i}{2^n \tilde{L}_{n+1}}$, $c = c_n$ is chosen such that

$$\varepsilon_n > C \cdot \tilde{L}_{n+1} \cdot n \cdot P, \tag{4}$$

K is chosen to satisfy

$$0 < 1 - \left[\varepsilon_n + c \sum_{k=1}^{K-2} (k+1)^{-1}\right] < \varepsilon_n,$$

$$\alpha_{n,o} = \varepsilon_n,\ \alpha_{n,k} = C\ (k+1)^{-1} \quad (1\le k\le K-2),$$

$$\alpha_{n,K-1} = 1 - \sum_{k=0}^{K-2} \alpha_{n,k}.$$

Set $\tilde{G}_{i,k} = \sigma^{i-1} \tilde{G}_{1,k}$, $1\le i\le P$, $0\le k\le K-1$,

and, therefore, represent $\tilde{\Omega}_{n+1}$ as a disjoint union of towers $\tilde{G}^{(k)}$, $0\le k\le K-1$, where $\tilde{G}^{(k)} = \bigcup_{i=1}^{P} G_{i,k}$.

e) Now we are going to construct the (n+1)-th tower Ω_{n+1} out of towers $\tilde{G}^{(k)}$ by putting each $\tilde{G}^{(k)}$ on top of the previous one. In order to do this, we will define the new "gluing maps" π_k: $\tilde{G}_{P,k}$ $\tilde{G}_{1,k+1}$, $0\le k\le K-2$, and then we will set $\sigma|_{\tilde{G}_{P,k}} = \pi_k$.

Only at this point do we need the restriction that τ is of type III_λ, $\lambda\neq 0$. The following lemma is a slight modification of a lemma in [5], p. 166, which is, in turn, based on Lemma 1.

LEMMA 4. Let τ be an automorphism of type III_λ, $\lambda\neq 0$, of $(\Omega,\mathcal{B},\mu)$, and $A,B,C\in\mathcal{B}$ are of positive measure. For any $\varepsilon>0$ and any transformation $\sigma_1\in[\tau]$ for which $\sigma_1 A = B$ there exists a transformation $\sigma_2\in[\tau]$ for which $\sigma_2 B = C$ such that $\sigma = \sigma_1 o \sigma_2$, $\sigma A = C$, satisfies

$$(\lambda-\varepsilon)\,\frac{\mu(C)}{\mu(A)} \le \frac{d\sigma\mu}{d\mu}(x) \le \left[\frac{1}{\lambda}+\varepsilon\right]\frac{\mu(C)}{\mu(A)}$$

for μ-a.e. $x\varepsilon A$.

We omit the proof of this lemma which is based on essentially the same ideas as the proof of Lemma 1.

To construct the map π_k, $k\ge 1$, we apply Lemma 4 to $A = \tilde{G}_{1,1}$, $B = \tilde{G}_{p,k}$, $C = \tilde{G}_{1,k+1}$ (and $\varepsilon = \frac{\lambda}{2}$). The map σ_2 from Lemma 4 will be our π_k. The map π_0 is obtained by applying Lemma 4 to $A = \tilde{G}_{1,0}$, $B = \tilde{G}_{P,0}$, $C = \tilde{G}_{1,1}$.

Let us prove now vi_{n+1}). Fix a point $x\in\tilde{G}_{i,k}$, $k>0$. Note that $\tilde{G}_{i,k} = \sigma^{\ell}\tilde{G}_{1,0}$, where $\ell = k\cdot p+i$. Since $f(x) = 1$ on $\tilde{G}_{1,0}$, we have

$$(S^{\ell}f)(x) = \frac{d\sigma^{-\ell}\mu}{d\mu}(x). \tag{5}$$

Let $y = \sigma^{-i}x$, $z = \sigma^{-(k-1)P}y$. Then $y\in\tilde{G}_{1,k}$, $z\in\tilde{G}_{1,1}$, and

$$\frac{d\sigma^{-\ell}\mu}{d\mu}(x) = \frac{d\sigma^{-\ell}\mu}{d\sigma^{-kP}\mu}(x)\cdot\frac{d\sigma^{-kP}\mu}{d\sigma^{-P}\mu}(y)\cdot\frac{d\sigma^{-P}\mu}{d\mu}(z).$$

From (3) we have:

$$\frac{d\sigma^{-\ell}\mu}{d\sigma^{-kP}\mu}(x) \ge \frac{1}{\tilde{L}} \tag{6}$$

Lemma 4 gives us

$$\frac{d\sigma^{-kP}\mu}{d\sigma^{-P}\mu}(y) \ge \text{const.}\left[\frac{\mu(\tilde{G}_{1,k})}{\mu\tilde{G}_{1,1})}\right]^{-1} \ge \text{const.}\ k. \tag{7}$$

Finally, from the construction of the map π_1:

$$\frac{d\sigma^{-P}\mu}{d\mu}(z) \geq \text{const.} \left[\frac{\mu(\tilde{G}_{1,1})}{\mu(\tilde{G}_{1,0})}\right]^{-1} \geq \text{const.} \frac{\varepsilon_n}{C}. \tag{8}$$

Combining (6), (7), (8) and taking into account (4), we have:

$$\frac{d\sigma^{-\ell}\mu}{d\mu}(x) \geq \text{const}\cdot \frac{k}{\tilde{L}} \cdot \tilde{L} \cdot nP \geq \text{const} \cdot n \cdot \ell.$$

Now (5) gives us vi_{n+1}, and this completes the proof of the theorem.

I would like to thank D. Cömez for useful discussions.

This research was supported by a grant from the Office of the Dean, Graduate Program and Research, North Dakota State University, whose assistance is greatly appreciated.

REFERENCES

1. R. V. Chacon. A class of linear transformations. Proc. Amer. Math. Soc., 15 (1964), 560-564.

2. J. R. Choksi, J. M. Hawkins, V. S. Prasad. Abelian cocycles for non-singular ergodic transformations and the genericity of type III_1 transformations. Mh. Math., 103 (1987), 187-205.

3. A. Ionescu Tulcea. On the category of certain classes of transformations in ergodic theory. Trans. Amer. Math. Soc., 114 (1965) 261-279.

4. U. Krengel. Ergodic Theorems. de Gruyter, Berlin, 1985.

5. W. Krieger. On the Araki-Woods asymptotic ratio set and non-singular transformations of a measure space. Lecture Notes in Math, Springer, 160 (1970), 158-177.

6. W. Krieger. On non-singular transformations of a measure space I, II. Z. Wahrscheinlichtkeitstheorie verw. Geb., 11 (1969), 83-119.

7. C. E. Sutherland. Notes on orbit equivalence: Krieger's theorem. Lecture Notes Series, Math. Institut, Universiteti Oslo, 1976.

8. H. Weiner. Invariant measures and Cesaro summability. Pacific J. Math., 25 (1968), 621-629.

9. Y. Ito. Invariant measures and pointwise ergodic theorem. Comment. Math. Univ. Sancti Paulu, 30 (1981), 193-201.

SOME ALMOST SURE CONVERGENCE PROPERTIES OF WEIGHTED SUMS OF MARTINGALE DIFFERENCE SEQUENCES

Tze Leung Lai

Department of Statistics, Stanford University

Stanford, CA 94305, USA

1. Introduction

Let $(\Omega, \mathcal{F}, P)$ be a probability space and let $(\mathcal{F}_n)_{n\geq 0}$ be a nondecreasing sequence of sub-σ-fields of $\mathcal{F}$. Let X_n be $q \times 1$ random vectors such that $(X_n)_{n\geq 1}$ is a martingale difference sequence with respect to the filtration $(\mathcal{F}_n)_{n\geq 0}$ (i.e., $E\|X_n\| < \infty$, X_n is $\mathcal{F}_n$-measurable and $E(X_n|\mathcal{F}_{n-1}) = 0$ for every n), and such that

$$\sup_n E(\|X_n\|^2|\mathcal{F}_{n-1}) < \infty \quad \text{a.s.} \tag{1.1}$$

Let $\{A_{ni} : n \geq 1, i \leq n\}$ be a triangular array of nonrandom $p \times q$ matrices. Weighted sums of the form $\sum_{i=1}^n A_{ni}X_i$ arise in a variety of applications. In Section 3 we review and generalize some almost sure convergence results for such weighted sums, which need not be martingales because of the double array structure of the weights A_{ni}.

In Section 2 we consider the case $A_{ni} = \Phi_i$ (not depending on n) but allow Φ_i to be random, requiring only that Φ_i be $\mathcal{F}_{i-1}$- measurable so that $(\sum_1^n \Phi_i X_i)_{n\geq 1}$ is a vector-valued martingale. Our study of this case has been motivated by multivariate stochastic regression models of the form

$$Y_i = \Phi_i' B + X_i, \quad i = 1, 2, \cdots, \tag{1.2}$$

where B is a $p \times 1$ vector of unknown parameters, Y_i represents the output at stage i corresponding to an input matrix Φ_i which may depend on the past observations (and is therefore $\mathcal{F}_{i-1}$-measurable), and the X_i represent unobservable random disturbances. The least squares estimate of B based on the observations $X_1, Y_1, \cdots, X_n, Y_n$ is

$$\hat{B}_n = (\sum_{i=1}^n \Phi_i\Phi_i')^{-1} \sum_{i=1}^n \Phi_i Y_i = B + (\sum_{i=1}^n \Phi_i\Phi_i')^{-1} \sum_{i=1}^n \Phi_i X_i, \tag{1.3}$$

ISBN 0-12-085520-8

assuming $\sum_1^n \Phi_i \Phi_i'$ to be nonsingular. In the scalar case $p = q = 1$, it is well known that under (1.1),

$$\begin{aligned} &\sum_{i=1}^n \Phi_i X_i \quad \text{converges a.s. on} \quad \{\sum_1^\infty \Phi_i^2 < \infty\} \quad \text{and} \\ &(\sum_{i=1}^n \Phi_i X_i)/(\sum_1^n \Phi_i^2) \to 0 \quad \text{a.s. on} \quad \{\sum_1^\infty \Phi_i^2 = \infty\}. \end{aligned} \tag{1.4}$$

An obvious multivariate generalization of this result is to replace Φ_i^2 in (1.4) by $\|\Phi_i\|^2$. An alternative extension, which is related to the a.s. convergence of $\hat{B}_n$ to B in view of (1.3), is to replace the norming factor $1/(\sum_1^n \Phi_i^2)$ in (1.4) by the matrix $(\sum_1^n \Phi_i \Phi_i')^{-1}$, leading to the "matrix-normed" strong law

$$(\sum_{i=1}^n \Phi_i \Phi_i')^{-1} \sum_{i=1}^n \Phi_i X_i \to 0 \quad \text{a.s. on} \quad \{\lambda_{\min}(\sum_{i=1}^n \Phi_i \Phi_i') \to \infty\}. \tag{1.5}$$

Here and in the sequel we let $\lambda_{\min}(\Gamma)$ and $\lambda_{\max}(\Gamma)$ denote the minimum and maximum eigenvalues, respectively, of a symmetric matrix Γ. Note that $\lambda_{\min}(\sum_{i=1}^n \Phi_i \Phi_i') \to \infty$ is equivalent to $(\sum_{i=1}^n \Phi_i \Phi_i')^{-1} \to 0$. It will be shown in Section 3 that (1.5) indeed holds when Φ_i is nonrandom. However, if Φ_i is a random $\mathcal{F}_{i-1}$-measurable matrix, (1.5) may fail to hold and some additional assumptions are required, as will be discussed in Section 2.

2. Matrix-normed strong laws

Throughout this section we shall assume that Φ_i is a $p \times q$ matrix that is $\mathcal{F}_{i-1}$-measurable. Let $\Gamma_n = \sum_{i=1}^n \Phi_i \Phi_i'$ and let $(X_n)_{n \geq 1}$ be a martingale difference sequence with respect to $(\mathcal{F}_n)_{n \geq 0}$ such that (1.1) holds.

THEOREM 1. (i) *$\sum_1^n \Phi_i X_i$ converges a.s. on* $\{\lim_{n\to\infty} \lambda_{\max}(\Gamma_n) < \infty\}$.
(ii) $\Gamma_n^{-1}(\sum_{i=1}^n \Phi_i X_i) \to 0$ *a.s. on* $\bigcup_{\rho>1} \{\lim_{n\to\infty} \lambda_{\min}(\Gamma_n)/[1 \vee \log \lambda_{\max}(\Gamma_n)]^\rho = \infty\}$.
(iii) *Suppose that* (1.1) *is strengthened into*

$$\sup_n E(\|X_n\|^r | \mathcal{F}_{n-1}) < \infty \quad a.s.\ for\ some \quad r > 2. \tag{2.1}$$

Then $\Gamma_n^{-1}(\sum_{i=1}^n \Phi_i X_i) \to 0$ *a.s. on* $\{\lim_{n\to\infty} \lambda_{\min}(\Gamma_n)/[1 \vee \log \lambda_{\max}(\Gamma_n)] = \infty\}$.

Theorem 1 is a generalization of Corollary 2 and Theorem 1 of Lai and Wei (1982a) who consider the special case $q = 1 \leq p$. The following counter-example of Lai and Wei (1982a), pages 159-160, shows that if $\lambda_{\min}(\Gamma_n) \to \infty$ and $\lim_{n\to\infty}(\log \lambda_{\max}(\Gamma_n))/\lambda_{\min}(\Gamma_n) \neq 0$ then $\Gamma_n^{-1}(\sum_1^n \Phi_i X_i)$ may fail to converge to 0.

EXAMPLE 1. Let $X_1, X_2, \cdots$ be i.i.d. random variables such that $EX_i = 0$ and $EX_i^2 = 1$. Let $U_n = \sum_{i=1}^{n-1} i^{-1} X_i$ ($U_1 = 0$), $\Phi_n = (1, U_n)'$, and let $\mathcal{F}_n$ be the σ-field generated by $X_1, \cdots, X_n$. Then Φ_i is $\mathcal{F}_{i-1}$-measurable and

$$\lambda_{\max}(\sum_1^n \Phi_i \Phi_i') \sim n\{1 + (\sum_1^\infty i^{-1} X_i)^2\} \quad \text{a.s.} \quad ,$$

$$\lambda_{\min}(\sum_1^n \Phi_i \Phi_i') \sim (\log n)/\{1 + (\sum_1^\infty i^{-1} X_i)^2\} \quad \text{a.s.}$$

Therefore $\lim_{n\to\infty}(\log \lambda_{\max}(\Gamma_n))/\lambda_{\min}(\Gamma_n) = 1 + (\sum_1^\infty i^{-1} X_i)^2$ a.s. Moreover,

$$\Gamma_n^{-1} \sum_1^n \Phi_i X_i \to (\sum_1^\infty i^{-1} X_i,\ -1)' \quad \text{a.s.}$$

Hence the additional condition $(\log \lambda_{\max}(\Gamma_n))/\lambda_{\min}(\Gamma_n) \to 0$ in Theorem 1(iii) is in some sense best possible for the matrix-normed strong law (1.5) to hold.

Theorem 1 can be proved by a modification of the arguments in Lai and Wei (1982a), pages 156-158. The basic idea is to apply martingale theory to analyze the recursive relations for the Lyapunov-type function

$$Q_n = (\sum_{i=1}^n \Phi_i X_i)' \Gamma_n^{-1} (\sum_{i=1}^n \Phi_i X_i) \tag{2.2}$$

associated with the recursions $\sum_1^n \Phi_i X_i = \sum_1^{n-1} \Phi_i X_i + \Phi_n X_n$ and $\Gamma_n = \Gamma_{n-1} + \Phi_n \Phi_n'$. Letting $N = \inf\{m : \lambda_{\min}(\Gamma_m) > 0\}$, note that $\lambda_{\min}(\Gamma_m)$ is nondecreasing in m and that for $n > N$, $\Gamma_n^{-1} = \Gamma_{n-1}^{-1} - \Gamma_{n-1}^{-1}\Phi_n(I + \Phi_n'\Gamma_{n-1}^{-1}\Phi_n)^{-1}\Phi_n'\Gamma_{n-1}^{-1}$, $\Gamma_n^{-1}\Phi_n = \Gamma_{n-1}^{-1}\Phi_n(I + \Phi_n'\Gamma_{n-1}^{-1}\Phi_n)^{-1}$. Therefore for $n > N$, Q_n satisfies the recursion

$$\begin{aligned} Q_n - Q_{n-1} = &-(\sum_{i=1}^{n-1} \Phi_i X_i)'\Gamma_{n-1}^{-1}\Phi_n(I + \Phi_n'\Gamma_{n-1}^{-1}\Phi_n)^{-1}\Phi_n'\Gamma_{n-1}^{-1}(\sum_{i=1}^{n-1} \Phi_i X_i) \\ &+ 2(\sum_{i=1}^{n-1} \Phi_i X_i)'\Gamma_{n-1}^{-1}\Phi_n(I + \Phi_n'\Gamma_{n-1}^{-1}\Phi_n)^{-1}X_n + X_n'\Phi_n'\Gamma_n^{-1}\Phi_n X_n. \end{aligned} \tag{2.3}$$

Summing the recursion (2.3) and applying Chow's (1965) martingale strong laws (cf. Lemma 2(iii) of Lai and Wei,1982a) to the martingale that corresponds to the second term on the right hand side of (2.3) gives

$$Q_m = -(1+o(1)) \sum_{n=N+1}^{m} \{(\sum_{i=1}^{n-1} \Phi_i X_i)' \Gamma_{n-1}^{-1} \Phi_n (I + \Phi_n' \Gamma_{n-1}^{-1} \Phi_n)^{-1} \Phi_n' \Gamma_{n-1}^{-1} (\sum_{i=1}^{n-1} \Phi_i X_i)\}$$
$$+ \sum_{n=N+1}^{m} X_n' \Phi_n' \Gamma_n^{-1} \Phi_n X_n + O(1) \quad \text{a.s.} \tag{2.4}$$

A simple modification of Lemma 2 of Lai and Wei (1982a) can be used to show that under the assumption (1.1),

$$\sum_{n=N+1}^{m} X_n' \Phi_n' \Gamma_n^{-1} \Phi_n X_n = O((\log \lambda_{\max}(\Gamma_m))^\rho) \quad \text{a.s.} \tag{2.5}$$

for every $\rho > 1$, and that we can set $\rho = 1$ in (2.5) under the stronger assumption (2.1).

Combining (2.4) and (2.5) gives that $Q_n = O((\log \lambda_{\max}(\Gamma_n))^\rho)$ a.s. for every $\rho > 1$, and also for $\rho = 1$ under (2.1). Let $\|\Gamma\| = \sup_{\|x\|=1} \|\Gamma x\|$ for a $p \times p$ matrix Γ. Since

$$\|\Gamma_n^{-1} \sum_{i=1}^{n} \Phi_i X_i\|^2 \le \|\Gamma_n^{-1/2}\|^2 \|\Gamma_n^{-1/2} \sum_{i=1}^{n} \Phi_i X_i\|^2 = (\lambda_{\min}(\Gamma_n))^{-1} Q_n,$$

we obtain Theorem 1(ii) and (iii).

The fact that the first term on the right hand side of the recursion (2.3) for Q_n is ≤ 0 plays a key role in the preceding argument. This property is analogous to that of Lyapunov functions commonly used in the stability analysis of differential (difference) equations. We first use it (upon summing the recursion) to dominate the martingale corresponding to the second term on the right hand side of (2.3) and then completely discard it to obtain an upper bound for Q_m. Since $Q_m \ge 0$, (2.4) and (2.5) also give the following result for the first term on the right hand side of (2.4).

THEOREM 2. *On* $\{N < \infty$ *and* $\lim_{m\to\infty} \lambda_{\max}(\Gamma_m) = \infty\}$,

$$\sum_{n=N+1}^{m} \|(I + \Phi_n' \Gamma_{n-1}^{-1} \Phi_n)^{-1/2} \Phi_n' \Gamma_{n-1}^{-1} (\sum_{i=1}^{n-1} \Phi_i X_i)\|^2 = O((\log \lambda_{\max}(\Gamma_m))^\rho) \quad a.s.$$

for every $\rho > 1$ under the assumption (1.1), *and for $\rho = 1$ under the stronger assumption* (2.1).

Applying Theorem 2 to the least squares estimate $\hat{B}_n$ defined in (1.3) gives

$$\begin{aligned}\sum_{n=N+1}^{m} \|\Phi_n'(\hat{B}_{n-1} - B)\|^2 = O((\log \lambda_{\max}(\Gamma_m))^\rho) \quad \text{a.s. on} \\ \{N < \infty, \lim_{n\to\infty} \Phi_n'\Gamma_{n-1}^{-1}\Phi_n = 0, \lim_{n\to\infty} \lambda_{\max}(\Gamma_n) = \infty\},\end{aligned} \tag{2.6}$$

for every $\rho > 1$ under (1.1), and for $\rho = 1$ under (2.1). Note that if B is known, then $E(Y_n|\mathcal{F}_{n-1}) = \Phi_n' B$ is the minimum variance predictor of the output Y_n based on observations up to stage $n-1$ in the stochastic regression model (1.2). If B is unknown, it is natural to "adapt" this predictor by replacing B by the least squares estimate $\hat{B}_{n-1}$, leading to the adaptive predictor $\Phi_n'\hat{B}_{n-1}$. The left hand side of (2.6) therefore measures the cumulative squared difference, up to stage m, between the optimal and adaptive predictors. This kind of results has important implications on adaptive control problems concerning the choice of the inputs Φ_i to drive the outputs Y_i towards certain target values y_i^* in ignorance of the parameter vector B, cf. Lai (1986).

3. Triangular arrays of nonrandom weights

Although Example 1 shows that the minimal assumption $(\sum_1^n \Phi_i\Phi_i')^{-1} \to 0$ may be insufficient to ensure the matrix-normed strong law $(\sum_1^n \Phi_i\Phi_i')^{-1}\sum_1^n \Phi_i X_i \to 0$ a.s. in the case of predictable (i.e., $\mathcal{F}_{i-1}$-measurable) Φ_i, it turns out to be sufficient if the Φ_i are nonrandom. For the case $q = 1 \le p$, (1.5) has been shown to hold for nonrandom Φ_i by Lai, Robbins and Wei (1979). The basic ideas of their proof are (i) to make use of the property that

$$\sum_1^\infty c_i X_i \quad \text{converges a.s. for all nonrandom} \quad c_i \quad \text{with} \quad \sum_1^\infty c_i^2 < \infty, \tag{3.1}$$

(ii) to show that for all *nonrandom* $\nu \times 1$ vectors $\mathbf{u}_i$ such that $\sum_1^m \mathbf{u}_i\mathbf{u}_i'$ is positive definite for some m and for all *nonrandom* constants c_i,

$$\sum_{m+1}^\infty c_i^2\{1+\mathbf{u}_i'(\sum_1^{i-1}\mathbf{u}_j\mathbf{u}_j')^{-1}\mathbf{u}_i\} < \infty \Rightarrow \sum_{i=m+1}^n c_i\mathbf{u}_i'(\sum_{j=1}^{i-1}\mathbf{u}_j\mathbf{u}_j')^{-1}(\sum_{j=1}^{i-1}\mathbf{u}_j X_j) \ \text{converges a.s.,}$$

and (iii) to represent any fixed component of the $p \times 1$ vector $(\sum_1^n \Phi_i \Phi_i')^{-1} \sum_1^n \Phi_i X_i$ in the form

$$\{\sum_{i=1}^n d_i [X_i - \mathbf{u}_i'(\sum_1^{i-1} \mathbf{u}_j \mathbf{u}_j')^{-1}(\sum_1^{i-1} \mathbf{u}_j X_j)]\} / \{\sum_{i=1}^n d_i^2 [1 + \mathbf{u}_i'(\sum_1^{i-1} \mathbf{u}_j \mathbf{u}_j)^{-1} \mathbf{u}_i]\}, \quad (3.2)$$

for some nonrandom scalars d_i and $(p-1) \times 1$ vectors $\mathbf{u}_i$.

For the case of general q, let ϕ_{ij} $(j = 1, \cdots, q)$ denote the column vectors of $\Phi_i (= (\phi_{i1}, \cdots, \phi_{iq}))$ and let $X_i = (X_{i1}, \cdots, X_{iq})'$. Let $\mathbf{u}_{(i-1)q+j} = \phi_{ij}$ and $\epsilon_{(i-1)q+j} = X_{ij}$ $(j = 1, \cdots, q; i \geq 1)$. Then

$$(\sum_{i=1}^n \Phi_i \Phi_i')^{-1} (\sum_{i=1}^n \Phi_i X_i) = (\sum_{i=1}^n \sum_{j=1}^q \phi_{ij} \phi_{ij}')^{-1} (\sum_{i=1}^n \sum_{j=1}^q \phi_{ij} X_{ij}) = (\sum_{t=1}^{nq} \mathbf{u}_t \mathbf{u}_t')^{-1} \sum_{t=1}^{nq} \mathbf{u}_t \epsilon_t .$$

Noting that $\sum_1^n c_i \epsilon_i$ converges a.s. for all nonrandom c_i with $\sum_1^\infty c_i^2 < \infty$, we can therefore apply the result of Lai, Robbins and Wei (1979) for the case $q = 1$ to obtain

THEOREM 3. *Suppose that* (1.1) *holds and that the* Φ_n *are nonrandom* $p \times q$ *matrices such that* $\lim_{n \to \infty} \lambda_{\min}(\sum_{i=1}^n \Phi_i \Phi_i') = \infty$. *Then* $(\sum_1^n \Phi_i \Phi_i')^{-1} \sum_1^n \Phi_i X_i \to 0$ *a.s.*

A law of the iterated logarithm for a fixed component (3.2) of $(\sum_1^n \Phi_i \Phi_i')^{-1} \sum_1^n \Phi_i X_i$ in the case of nonrandom $p \times 1$ vectors Φ_i has been established by Lai and Wei (1982b) under the assumption that the X_n are independent random variables with $EX_n = 0$, $EX_n^2 = \sigma^2$ and $\sup_n E|X_n|^r < \infty$ for some $r > 2$. Their proof depends on the following dual structure of the numerator of (3.2). Let $Z_t = X_t - \mathbf{u}_t'(\sum_1^{t-1} \mathbf{u}_j \mathbf{u}_j')^{-1} \sum_1^{t-1} \mathbf{u}_j X_j$. Then $E(Z_t Z_s) = 0$ for $t > s$ and therefore the numerator of (3.2) is a weighted sum $\sum_1^n d_i Z_i$ of the orthogonal random variables Z_i with a single array of weights d_i. On the other hand, we can also express the numerator of (3.2) as a weighted sum $\sum_{i=1}^n a_{ni} X_i$ of the independent random variables X_i with a double array of weights $a_{ni} = d_i - \sum_{i+1 \leq t \leq n} d_t \mathbf{u}_t'(\sum_{j=1}^{t-1} \mathbf{u}_j \mathbf{u}_j')^{-1} \mathbf{u}_i$. Taking $\sigma = 1$ shows that for $n > m$

$$\sum_{i=m+1}^n a_{ni}^2 + \sum_{i=1}^m (a_{ni} - a_{mi})^2 = \operatorname{Var}(\sum_{i=1}^n a_{ni} X_i - \sum_{i=1}^m a_{mi} X_i) = \operatorname{Var}(\sum_{i=m+1}^n d_i Z_i)$$

$$= \sum_{i=m+1}^n d_i^2 \operatorname{Var}(Z_i) = \sum_{i=m+1}^n d_i^2 \{1 + \mathbf{u}_i'(\sum_{j=1}^{i-1} \mathbf{u}_j \mathbf{u}_j')^{-1} \mathbf{u}_i\}.$$

Hence these weights a_{ni} satisfy assumptions (3.4) and (3.5) of the following theorem, which is an extension of Theorem 1 of Lai and Wei (1982b) that assumes independent univariate X_n to multivariate martingale differences X_n. Moreover, under the assumptions $\Gamma_{n+1} = O(\Gamma_n)$ and (3.3) below on these weights, it has been shown in the proof of Theorem 2 of Lai and Wei (1982b) how an increasing sequence of positive integers $n_i(= n_i(\gamma))$ satisfying condition (3.9) of the following theorem can be constructed for every $0 < \gamma < 1$. Since different components of $(\sum_1^n \Phi_i\Phi_i')^{-1}\sum_1^n \Phi_i X_i$ may have very different norming factors in the law of iterated logarithm, we treat these components separately instead of using the largest norming factor to normalize the entire vector.

THEOREM 4. *Suppose that the martingale difference sequence of $q \times 1$ vectors $(X_n)_{n\geq 1}$ satisfies (1.1) and $\sup_n E\|X_n\|^r < \infty$ for some $r > 2$. Let $\{A_{ni} : 1 \leq i \leq n, n \geq 1\}$ be a triangular array of $1 \times q$ nonrandom vectors and B be a symmetric positive definite $q \times q$ nonrandom matrix such that as $n \to \infty$,*

$$\Gamma_n = \sum_{i=1}^n A_{ni}BA'_{ni} \to \infty \text{ and } \max_{i\leq n}\|A_{ni}\|^2 = o(\Gamma_n(\log\Gamma_n)^{-\rho}) \text{ for all } \rho > 0. \tag{3.3}$$

(i) *Suppose there exist constants $c_i \geq 0$ and $d > 2/r$ such that*

$$\sum_{i=m+1}^n \|A_{ni}\|^2 + \sum_{i=1}^m \|A_{ni} - A_{mi}\|^2 \leq (\sum_{i=m+1}^n c_i)^d \text{ for } n > m \geq m_0, \tag{3.4}$$

$$(\sum_{i=m_0}^n c_i)^d = O(\Gamma_n) \text{ as } n \to \infty. \tag{3.5}$$

Then

$$\begin{aligned}&\limsup_{n\to\infty}|\sum_{i=1}^n A_{ni}X_i|/(2\Gamma_n\log\log\Gamma_n)^{1/2}\\ &\qquad \leq \limsup_{i\to\infty}\lambda_{\max}^{1/2}(B^{-1/2}E[X_iX_i'|\mathcal{F}_{i-1}]B^{-1/2}) \text{ a.s.}\end{aligned} \tag{3.6}$$

(ii) *Suppose that $\sum_{i=m+1}^n \|A_{ni}\|^2 + \sum_{i=1}^m \|A_{ni} - A_{mi}\|^2 \leq g(n-m)$ for $n > m \geq m_0$, where g is a positive function on $\{1, 2, \cdots\}$ such that $g(n) = O(\Gamma_n)$ and*

$$\begin{aligned}&\liminf_{n\to\infty} g(Kn)/g(n) > K^{2/r} \text{ for some integer } K \geq 2,\\ &\forall\gamma > 0, \exists\delta < 1 \text{ such that } \limsup_{n\to\infty}\{\max_{\delta n\leq i\leq n} g(i)/g(n)\} < 1 + \gamma.\end{aligned} \tag{3.7}$$

Then (3.6) *still holds.*

(iii) *Assume instead of* (1.1) *the stronger condition*

$$\lim_{i\to\infty} E(X_iX_i'|\mathcal{F}_{i-1}) = B \ a.s. \tag{3.8}$$

Suppose that for every $0 < \gamma < \gamma_0$ *there exist integers* $1 < n_1 < n_2 < \cdots$ *(depending on* γ*) such that*

$$\begin{aligned}&\limsup_{k\to\infty}\Big(\sum_{i=1}^{n_{k-1}} \|A_{n_k,i}\|^2\Big)//\Gamma_{n_k} \le \gamma, \quad \liminf_{k\to\infty}(\log\log\Gamma_{n_k})/\log k > 0,\\ &\limsup_{k\to\infty}(\log\log\Gamma_{n_k})/(\log k) \le 1+\gamma.\end{aligned} \tag{3.9}$$

Then $\liminf_{n\to\infty} |(2\Gamma_n\log\log\Gamma_n)^{-1/2}\sum_{i=1}^n A_{ni}X_i - \beta| = 0$ *a.s. for every* $\beta\in[-1,1]$.

PROOF. Let $S_n = \sum_{i=1}^n A_{ni}X_i$. Since $A_{ni}X_i = (A_{ni}B^{1/2})(B^{-1/2}X_i)$ and since $\|AB^{1/2}\|^2 \le \|A\|^2\lambda_{\max}(B)$, we shall assume, without loss of generality (upon replacing A_{ni} by $A_{ni}B^{1/2}$ and X_i by $B^{-1/2}X_i$), that $B = I$. Using Burkholder's (1973) inequality and Minkowski's inequality, it can be shown that for $n > m$

$$E|S_n - S_m|^r \le C_r(\sup_i E\|X_i\|^r)\Big\{\sum_{i=m+1}^n \|A_{ni}\|^2 + \sum_{i=1}^m \|A_{ni} - A_{mi}\|^2\Big\}^{r/2}, \tag{3.10}$$

where $C_r > 0$ depends only on r, cf. Lemma 2.1 of Lai and Wei (1983).

We next show that for every $\xi > 1$, $\theta > 0$, $k \ge 1$ and $v > 0$,

$$P\{|S_n| > \xi(2\theta v\Gamma_n\log\log\Gamma_n)^{1/2}, v > \max_{k\le i\le n}\lambda_{\max}(E(X_iX_i'|\mathcal{F}_{i-1}))\} = O(e^{-\theta\log\log\Gamma_n}). \tag{3.11}$$

From (3.3) and $\sup_i E\|X_i\|^r < \infty$, it follows as in (4.23) of Lai and Wei (1982b) that

$$P\{|A_{ni}X_i| \ge \Gamma_n^{1/2}(\log\log\Gamma_n)^{-1} \text{ for some } i\le n\} = o(\exp(-\theta\log\log\Gamma_n)). \tag{3.12}$$

Let $\epsilon_{ni} = X_iI_{\{|A_{ni}X_i|<\Gamma_n^{1/2}(\log\log\Gamma_n)^{-1}\}}$, $\tilde S_n = \sum_{i=k}^n A_{ni}\{\epsilon_{ni} - E(\epsilon_{ni}|\mathcal{F}_{i-1})\}$. Since $E(X_i|\mathcal{F}_{i-1}) = 0$, $|E(A_{ni}\epsilon_{ni}|\mathcal{F}_{i-1})| \le \{\Gamma_n^{1/2}(\log\log\Gamma_n)^{-1}\}^{-(r-1)}E(|A_{ni}X_i|^r|\mathcal{F}_{i-1})$. Hence

$$\begin{aligned}P\Big\{\sum_{i=1}^n |A_{ni}E(\epsilon_{ni}|\mathcal{F}_{i-1})| \ge \Gamma_n^{1/2}\Big\} &\le \Gamma_n^{-1/2}\sum_{i=1}^n E|E(A_{ni}\epsilon_{ni}|\mathcal{F}_{i-1})|\\ &\le \Gamma_n^{-r/2}(\log\log\Gamma_n)^{r-1}\sum_{i=1}^n \|A_{ni}\|^r E\|X_i\|^r\\ &\le (\sup_i E\|X_i\|^r)\Gamma_n^{-(r-2)/2}(\max_{i\le n}\|A_{ni}\|^{r-2})(\log\log\Gamma_n)^{r-1}\\ &= o(\exp(-\theta\log\log\Gamma_n)),\end{aligned} \tag{3.13}$$

by (3.3). Noting that $E((A_{ni}\epsilon_{ni})^2|\mathcal{F}_{i-1}) \le \|A_{ni}\|^2\lambda_{\max}(E(X_iX_i'|\mathcal{F}_{i-1}))$, application of the exponential inequality for martingales (cf. Lemma 1.1 of Stout, 1973) shows that (3.11) holds with $\tilde{S}_n$ in place of S_n. Combining this with (3.12) and (3.13) gives (3.11).

In view of (3.10) and (3.11), we can apply Theorem 4 of Lai and Wei (1982b, pages 328 and 330-331) to prove under the assumptions of (i) or (ii) that for every $\zeta > 1$, $v > 0$ and $k \ge 1$,

$$P\{|S_n| > \zeta(2v\Gamma_n \log\log\Gamma_n)^{1/2} \quad \text{and} \quad v > \max_{k\le i\le n} \lambda_{\max}(E(X_iX_i'|\mathcal{F}_{i-1})) \\ \text{for infinitely many } n's\} = 0.$$

Summing the above probabilities over positive rational $v's$ then shows that

$$P\{|S_n| > \zeta[2\max_{k\le i\le n} \lambda_{\max}(E(X_iX_i'|\mathcal{F}_{i-1}))\Gamma_n \log\log\Gamma_n]^{1/2} \text{ infinitely often }\} = 0,$$

for every $\zeta > 1$ and $k \ge 1$. Hence

$$\lim\ \sup_{n\to\infty}|S_n|/(2\Gamma_n \log\log\Gamma_n)^{1/2} \le \{\sup_{i\ge k} \lambda_{\max}(E(X_iX_i'|\mathcal{F}_{i-1})\}^{1/2} \quad \text{a.s.}$$

for every $k \ge 1$. Letting $k \to \infty$ then proves (i) and (ii), under the assumption $B = I$.

We now sketch the proof of (iii). Let $0 < \gamma < 1/5 \wedge \gamma_0$, and let $Z_1, Z_2, \cdots$ be i.i.d. bounded $q \times 1$ random vectors such that Z_n is independent of $\mathcal{F}_n$ and $EZ_n = 0$, $E(Z_nZ_n') = I$ for every n. Defining ϵ_{ni} as before, let

$$\begin{aligned} Y_{ni} &= \epsilon_{ni} - E(\epsilon_{ni}|\mathcal{F}_{i-1}) \quad \text{if} \quad \|\text{Cov}(\epsilon_{ni}|\mathcal{F}_{i-1}) - I\| \le \gamma, \\ &= Z_i \quad \text{otherwise}. \end{aligned} \tag{3.14}$$

Let $\mathcal{G}_n$ be the σ-field generated by $Z_1, \cdots, Z_n$ and $\mathcal{F}_n$. Since $\text{Cov}(\epsilon_{ni}|\mathcal{F}_{i-1})$ is $\mathcal{F}_{i-1}$-measurable, $\{Y_{ni}, \mathcal{G}_i, i \le n\}$ is a martingale such that $\|\text{Cov}(Y_{ni}|\mathcal{G}_{i-1}) - I\| \le \gamma$. An argument similar to the development of lower exponential bounds for martingales in Stout (1970), pages 283-286, can be used to show that for $0 < \eta \le 1$,

$$P\{\sum_{i=n_{k-1}+1}^{n_k} A_{n_k,i}Y_{n_k,i} \ge [2(1-4\gamma)\eta\Gamma_{n_k}\log\log\Gamma_{n_k}]^{1/2}|\mathcal{G}_{n_{k-1}}\} \\ \ge \exp\{-(1-\gamma)\eta\log\log\Gamma_{n_k}\},$$

for all large k, in view of (3.9). Moreover, an argument similar to the proof of Lemma 1 and Corollary 2 of Stout (1970) shows that for all large k

$$P\{\sum_{i=n_{k-1}+1}^{n_k} A_{n_k,i}Y_{n_k,i} > [2(\eta+\gamma)\Gamma_{n_k}\log\log\Gamma_{n_k}]^{1/2}|\mathcal{G}_{n_{k-1}}\}$$
$$\leq \exp(-\eta\log\log\Gamma_{n_k}) = o(\exp\{-(1-\gamma)\eta\log\log\Gamma_{n_k}\}).$$

Hence application of Lévy's (1937) conditional Borel-Cantelli lemma shows that

$$\begin{aligned} P\{(\eta+\gamma)^{1/2} \geq (2\Gamma_{n_k}\log\log\Gamma_{n_k})^{-1/2} \sum_{i=n_{k-1}+1}^{n_k} A_{n_k,i}Y_{n_k,i} \geq (1-4\gamma)^{1/2}\eta^{1/2} \\ \text{for infinitely many } k's\} = 1, \end{aligned} \tag{3.15}$$

noting that $\limsup_{k\to\infty}(\log\log\Gamma_{n_k})/\log k \leq 1+\gamma$.

Since (3.12) and (3.13) hold for all $\theta > 0$ and since $\liminf_{k\to\infty}(\log\log\Gamma_{n_k})/\log k > 0$, it follows from the Borel-Cantelli lemma that

$$\begin{aligned} P\{|\sum_{i=n_{k-1}+1}^{n_k} A_{n_k,i}X_i - \sum_{i=n_{k-1}+1}^{n_k} A_{n_k,i}\{\epsilon_{n_k,i} - E(\epsilon_{n_k,i}|\mathcal{F}_{i-1})\}| \geq \Gamma_{n_k}^{1/2} \\ \text{for infinitely many } k's\} = 0. \end{aligned} \tag{3.16}$$

Letting $D_{i,k} = \{\|\text{Cov}(X_i|\mathcal{F}_{i-1}) - I\| \leq \gamma/2, \|\text{Cov}(\epsilon_{n_k,i}|\mathcal{F}_{i-1}) - I\| > \gamma\}$, we shall show that

$$\begin{aligned} P\{|\sum_{i=n_{k-1}+1}^{n_k} A_{n_k,i}\{\epsilon_{n_k,i} - E(\epsilon_{n_k,i}|\mathcal{F}_{i-1}) - Z_i\}I_{D_{i,k}}| \geq \Gamma_{n_k}^{1/2} \\ \text{for infinitely many } k's\} = 0. \end{aligned} \tag{3.17}$$

Since $\lim_{i\to\infty}\text{Cov}(X_i|\mathcal{F}_{i-1}) = I$ a.s., it then follows from (3.14)-(3.17) that

$$\begin{aligned} P\{(\eta+2\gamma)^{1/2} \geq (2\Gamma_{n_k}\log\log\Gamma_{n_k})^{-1/2} \sum_{i=n_{k-1}+1}^{n_k} A_{n_k,i}X_i \geq (1-5\gamma)^{1/2}\eta^{1/2} \\ \text{for infinitely many } k's\} = 1. \end{aligned} \tag{3.18}$$

By (3.9), $\log\log\Gamma_{n_k} \geq \delta\log k$ for some $0 < \delta < 1$ and all large k. By an argument similar to the proof of (3.11) and the Borel-Cantelli lemma, it can be shown that

$$P\{|\sum_{i=1}^{n_{k-1}} A_{n_k,i}X_i| \geq (3\gamma\delta^{-1}\Gamma_{n_k}\log\log\Gamma_{n_k})^{1/2} \text{ infinitely often}\} = 0, \tag{3.19}$$

noting that $\limsup_{k\to\infty}(\sum_{i=1}^{n_{k-1}} \|A_{n_k,i}\|^2)/\Gamma_{n_k} \le \gamma$ by (3.9). Since $\gamma > 0$ can be chosen arbitrarily small, (3.18) and (3.19) give the desired conclusion (iii) with $\beta = \eta$, and replacing X_i by $-X_i$ in the preceding argument also shows that (iii) holds with $\beta = -\eta$.

To prove (3.17), note that Z_i is independent of $\mathcal{F}_i$, $D_{i,k}$ is $\mathcal{F}_{i-1}$-measurable and $\mathrm{Var}(A_{n_k,i}\epsilon_{n_k,i}|\mathcal{F}_{i-1}) \le \|A_{n_k,i}\|^2 E(\|\epsilon_{n_k,i}\|^2|\mathcal{F}_{i-1}) \le \|A_{n_k,i}\|^2 E(\|X_i^2\||\mathcal{F}_{i-1})$. Hence

$$
\begin{aligned}
P\{|\sum_{i=n_{k-1}+1}^{n_k} A_{n_k,i}\{\epsilon_{n_k,i} - E(\epsilon_{n_k,i}|\mathcal{F}_{i-1}) - Z_i\}I_{D_{i,k}}| \ge \Gamma_{n_k}^{1/2}\} \\
\le \Gamma_{n_k}^{-1} \sum_{i=n_{k-1}+1}^{n_k} \|A_{n_k,i}\|^2\{E[E(\|X_i\|^2|\mathcal{F}_{i-1})I_{D_{i,k}}] + P(D_{i,k})\} \qquad (3.20)\\
\le \max_{n_{k-1}<i\le n_k} \{(\sup_n E\|X_n\|^r)^{2/r}(P(D_{i,k}))^{(r-2)/r} + P(D_{i,k})\},
\end{aligned}
$$

by the Hölder inequality. On $D_{i,k}, \gamma/2 < \|\mathrm{Cov}(\epsilon_{n_k,i}|\mathcal{F}_{n-1}) - \mathrm{Cov}(X_i|\mathcal{F}_{i-1})\| \le 2E[\|X_i\|^2 I_{\{|A_{n_k,i}X_i|\ge\Gamma_{n_k}^{1/2}(\log\log\Gamma_{n_k})^{-1}\}}|\mathcal{F}_{i-1}]$, and therefore

$$
\begin{aligned}
\max_{i\le n_k} P(D_{i,k}) &\le (4/\gamma\}\max_{i\le n_k} E(\|X_i\|^2 I_{\{|A_{n_k,i}X_i|\ge\Gamma_{n_k}^{1/2}(\log\log\Gamma_{n_k})^{-1}\}}) \\
&\le (4/\gamma)(\sup_i E\|X_i\|^r)(\max_{i\le n_k}\|A_{n_k,i}\|/\Gamma_{n_k}^{1/2})^{r-2}(\log\log\Gamma_{n_k})^{r-2} \qquad (3.21)\\
&= o(\exp(-\theta\log k))
\end{aligned}
$$

for every $\theta > 0$, by (3.3) and (3.9). From (3.20) and (3.21), (3.17) follows by the Borel-Cantelli lemma, completing the proof of (iii).

EXAMPLE 2. Let $f : [0,1] \to R^q$ be Hölder continuous with exponent 1/2, i.e., $\|f(x) - f(y)\| \le K|x-y|^{1/2}$ for some $K > 0$ and all $x, y \in [0,1]$. Suppose that the martingale difference sequence $(X_n)_{n\ge1}$ satisfies (3.8) and $\sup_n E\|X_n\|^r < \infty$ for some $r > 2$. Then Theorem 4(ii) and (iii) can be applied to show that

$$
\limsup_{n\to\infty}|\sum_{i=1}^{n} f(i/n)X_i|/(2n\log\log n)^{1/2} = \left\{\int_0^1 f(t)Bf(t)'\,dt\right\}^{1/2} \quad \text{a.s.}, \qquad (3.22)
$$

cf. Lai and Wei (1982b), pages 334-335.

EXAMPLE 3. Let $(C_n)_{n\ge0}$ be a nonrandom sequence of $1\times q$ vectors such that $\sum_0^\infty \|C_n\| < \infty$, $\sum_0^\infty C_n \ne 0$ and let $Z_n = \sum_{i=1}^n C_{n-i}X_i$ be a linear process generated by the martingale difference sequence $(X_n)_{n\ge1}$ that satisfies (3.8) and the condition

$\sup_n E\|X_n\|^r < \infty$ for some $r > 2$. Then Theorem 4(ii) and (iii) can be applied (cf. Lai and Wei, 1982b, pages 326-327) to show that

$$\lim\ \sup_{n\to\infty} |\sum_{i=1}^{n} Z_i|/(2n \log\log n) = (\sum_{i=0}^{\infty} C_i)B(\sum_{i=0}^{\infty} C_i)' \quad \text{a.s.} \tag{3.23}$$

REFERENCES

Burkholder, D. L. (1973). Distribution function inequalities for martingales. *Ann. Probab.* **1**, 19-42.

Chow, Y. S. (1965). Local convergence of martingales and the law of large numbers. *Ann. Math. Statist.* **36**, 552-558.

Lai, T. L. (1986). Asymptotically efficient adaptive control in stochastic regression models. *Adv. Appl. Math.* **7**, 23-45.

Lai, T. L., Robbins, H., and Wei, C. Z. (1979). Strong consistency of least squares estimates in multiple regression II. *J. Multivariate Anal.* **9**, 343-36l.

Lai, T. L. and Wei, C. Z. (1982a). Least squares estimates in stochastic regression models with applications to identification and control of dynamic systems. *Ann. Statist.* **10**, 154-166.

Lai, T. L. and Wei, C. Z. (1982b). A law of the iterated logarithm for double arrays of independent random variables with applications to regression and time series models. *Ann. Probab.* **10**, 320-335.

Lai, T. L. and Wei, C. Z. (1983). Lacunary systems and generalized linear processes. *Stoch. Processes & Appl.* **14**, 187-199.

Lévy, P. (1937). "Théorie de l'addition des variables aléatoires". Gauthier-Villars, Paris.

Stout, W. F. (1970). A martingale analogue of Kolmogorov's law of the iterated logarithm. *Z. Wahrsch. verw. Gebiete* **15**, 279-290.

Stout, W. F. (1973). Maximal inequalities and the law of the iterated logarithm. *Ann. Probab.* **1**, 322-328.

Pointwise ergodic theorems for certain order preserving mappings in L^1 [†]

MICHAEL LIN
Department of Mathematics
Ben Gurion University of the Negev

RAINER WITTMANN*
Department of Mathematics
Princeton University

Abstract. For an order preserving map T in L^1 we study the "nonlinear sums" defined recurrently by $S_0 f = f$, $S_{n+1} f = f + T(S_n f)$, and the "averages" $A_n f = S_n f/(n+1)$. We prove a maximal ergodic lemma if T is norm decreasing in L^1, dominated estimates for $Mf = \sup A_n f$ when T is also L^∞ nonexpansive and order continuous. If moreover T is positively homogeneous and has no fixed points $\neq 0$ in L^p_+ (resp. L^p) and $1 < p < \infty$ then we can show that $\lim A_n f = 0$ a.e. and in L^p–norm for $f \in L^p_+$ (resp. $f \in L^p$). This theorem is applied to generalized measure preserving transformations in infinite measure spaces. In this case a.e. convergence holds also for $p = 1$. In the finite measure case we add the assumption that T is integral preserving and that the constant functions are the only fixed points and obtain a.e. convergence of $A_n f$ to $\int f$ for $f \in L \log L$.

1. INTRODUCTION

Order preserving mappings T in L^1 or in L^1_+ of a σ–finite measure μ, which are integral preserving and satisfy $T0 = 0$, can serve as models for random motions of matter, when the movement of the mass in a part B of the space is affected by the distribution of the mass in the complement B^c. Such operators are necessarily L^1 nonexpansive (see [3] or [8]). The following ergodic theorem was proved in Krengel and Lin [8] :

Let T be order preserving and nonexpansive in L^1 (or L^1_+), and assume that T decreases the L^∞–norm. Then for $f \in L^p(L^p_+)$, $1 < p < \infty$, the "linear" averages $n^{-1}\sum_{k=0}^{n-1} T^k f$ converge weakly in L^p. If μ is finite, the same is true for $p = 1$.

An example in [8] shows that strong convergence need not hold. Another example by Krengel [6] shows that pointwise convergence may fail even when norm convergence holds for any $1 \leq p < \infty$.

In ergodic theory, when T is linear, the sums $S_n f = \sum_{k=0}^{n} T^k f$ play an important role. They satisfy the recursive definition $S_0 f = f$, $S_{n+1} f = f + T(S_n f) = f + S_n(Tf)$. In the nonlinear case, however, the latter equality fails. While the sums $Z_n := \sum_{k=0}^{n} T^k f$ still satisfy $Z_0 = f$, $Z_{n+1} f = f + Z_n(Tf)$, they are usually no more equal to $S_n f$, defined inductively by $S_0 f = f$, $S_{n+1} = f + T(S_n f)$. The sequence $S_n f$ can be thought of as an "evolution" in discrete time by the "law" T with "constant input" f.

The purpose of this paper is to study the properties of $S_n f$ and $A_n f = S_n f/(n+1)$ for order preserving operators in L^1. We obtain the following theorem :

Let T be order preserving, order continuous, L^1 norm decreasing, L^∞ nonexpansive and positively homogeneous (see sect. 2 for the definitions). Let $1 < p < \infty$. If T has no invariant

[†] Most of this work was done while the second author was visiting Ben Gurion University.
*Heisenberg Fellow of Deutsche Forschungsgemeinschaft

ISBN 0-12-085520-8

functions $\neq 0$ *in* $L_+^p(L^p)$, *then for every* $f \in L_+^p(L^p)$ *we have* $A_n f \to 0$ a.e., $\sup A_n f \in L^p$ *and* $\|A_n f\|_p \to 0$.

This theorem can be applied to generalized measure preserving transformations in infinite measure spaces, and in particular to the "speed limit operators" of [8] from which the above mentioned counterexamples are taken. The reason for convergence of $A_n f$ in contrast to divergence of the Cesarò averages is the following : while the "linear" sums do not take into consideration the matter which has been accumulated before, the "nonlinear" sums do this very well. In the case of speed limit operators a "traffic jam" makes the nonlinear averages convergent.

2. Order preserving mappings in L^1

In the sequel $(E, \mathcal{B}, \mu)$ will be a σ–finite measure space. For a mapping T on a set of measurable functions $\mathcal{F}$ on E we frequently use the following properties :
T is said to be *order preserving* if

$$f \leq g \Longrightarrow Tf \leq Tg \quad (f, g \in \mathcal{F})$$

T is said to be *order continuous* if

$$\begin{matrix} f_n \uparrow f \\ f_n \downarrow f \end{matrix} \Longrightarrow \lim_{b \to \infty} Tf_n = Tf \text{ a.e.} \quad (\{f, f_n, n \in \mathbf{N}\} \subset \mathcal{F})$$

Note that T may be order continuous without being order preserving (e.g. $Tf = -f$).

For $1 \leq p \leq \infty$ we say that T is L^p *norm decreasing* if $\|Tf\|_p \leq \|f\|_p$ $(f \in \mathcal{F} \cap L^p)$, and L^p *nonexpansive* if $\|Tf - Tg\|_p \leq \|f - g\|_p$ $(f, g \in \mathcal{F} \cap L^p)$. Clearly, if $\mathcal{F} \subset L^p$ for a given $1 \leq p < \infty$ and if T is L^p nonexpansive, then T is order continuous.

T is said to be *positively integral preserving* if

$$\int f \, d\mu = \int Tf \, d\mu \quad (f \in \mathcal{F}, f \geq 0)$$

T is said to be *integral preserving* if the above property holds for all $f \in \mathcal{F}$.
If $\mathcal{F}$ is a cone then we say that T is *positively homogeneous* if $T(\alpha f) = \alpha Tf$ $(f \in \mathcal{F}, \alpha \geq 0)$.

Expressions involving measurable functions or sets have always to be understood in the a.e. sense. Apart from some distinguished places this will not be mentioned anymore.

Lemma 2.1. *For any order continuous,* L^1 *norm decreasing and* L^∞ *nonexpansive, order preserving mapping* T *on* $L^1(\mu)$ *we have*

(i) $$Tf - t \leq T(f - t)_+ \quad (f \in L^1(\mu), t \in [0, \infty[)$$

(ii) $$\int (Tf - t)_+ \, d\mu \leq \int (f - t)_+ d\mu \quad (f \in L^1(\mu), t \in [0, \infty[)$$

Proof: To prove (i) we assume first that $f \in L_+^1 \cap L^\infty$. Then $\|f - (f - t)_+\|_\infty \leq t$ implies $\|Tf - T(f - t)_+\|_\infty \leq t$. In particular,

$$Tf \leq T(f - t)_+ + t \quad (f \in L_+^1 \cap L^\infty)$$

If $f \in L^1(\mu)$ is arbitrary then we choose a sequence (f_n) in $L_+^1 \cap L^\infty$ such that $f_n \uparrow f_+$. Now (i) follows from

$$Tf - t \leq Tf_+ - t = \lim_{n \to \infty} Tf_n - t \leq \lim_{n \to \infty} T(f_n - t)_+ = T(f_+ - t)_+ = T(f - t)_+$$

Since T is L^1–norm decreasing (ii) follows from (i). ∎

The following lemma is implicit in Crandall and Tartar [3].

LEMMA 2.2. *For any L^1 nonexpansive and L^∞ norm decreasing, order preserving mapping T on $L^1(\mu)$ we have*

$$\int (Tf - t)_+ \, d\mu \le \int (f - t)_+ d\mu \quad (f \in L^1(\mu), t \in [0, \infty[)$$

PROOF: Since T is L^∞ norm decreasing and order preserving we have $T(f \wedge t) \le t$. Using L^1 nonexpansiveness the assertion follows now from

$$\int (Tf - t)_+ \, d\mu \le \int |Tf - T(f \wedge t)| \, d\mu \le \int |f - (f \wedge t)| \, d\mu = \int (f - t)_+ \, d\mu \quad \blacksquare$$

REMARK 2.3: As the proofs show, Lemma 2.1 and 2.2 apply also for mappings T on L^1_+ which need not necessarily be order preserving.

PROPOSITION 2.4. *Let T be a mapping on $L^1_+(\mu)$ such that*

$$\text{(i)} \qquad \int (Tf - t)_+ \, d\mu \le \int (f - t)_+ \, d\mu \quad (f \in L^1_+, \, t \in [0, \infty[)$$

Then we have

$$\text{(ii)} \qquad \|Tf\|_p \le \|f\|_p \quad (1 < p < \infty, f \in L^p \cap L^1_+)$$

PROOF (CF. GARSIA [4]): For an arbitrary measurable function $h \ge 0$ and $1 < p < \infty$ Fubini's theorem yields

$$\begin{aligned}
&\int_0^\infty t^{p-2} \int (h - t)_+(x) \, d\mu(x) \, dt \\
&= \int_0^\infty t^{p-2} \int_{\{h>t\}} h(x) \, d\mu(x) \, dt - \int_0^\infty t^{p-2} \int_{\{h>t\}} t \, d\mu(x) \, dt \\
&= \int h(x) \int_0^{h(x)} t^{p-2} \, dt \, d\mu(x) - \int \int_0^{h(x)} t^{p-1} \, dt \, d\mu(x) \\
&= \left(\frac{1}{p-1} - \frac{1}{p} \right) \int h^p(x) \, d\mu(x)
\end{aligned}$$

Applying this equality with $h = f$ and $h = Tf$ the assertion follows from (i) $\blacksquare$

The next corollary generalizes Proposition 4 of Crandall and Tartar [**3**] and Proposition 2.6 of Krengel and Lin [**8**].

COROLLARY 2.5. *Let T be a mapping on L^1_+ which is either order continuous, norm decreasing in L^1 and L^∞ nonexpansive, or nonexpansive in L^1 and L^∞-norm decreasing. Then T is L^p-norm decreasing for any $1 \le p \le \infty$.*

If T is defined on all of L^1, order preserving and the other assumptions are as above, then T is L^p norm decreasing for any $1 \le p \le \infty$.

PROOF: By Remark 2.3 the assumption of Proposition 2.4 is fulfilled, and the first part of the assertion follows immediately.

When T is defined on all of L^1, the operator $f \longrightarrow -T(-f)$ satisfies the same assumptions as T, and we get

$$\|Tf_+\|_p \leq \|f_+\|_p, \; \|-T(-f_-)\|_p \leq \|f_-\|_p \quad (f \in L^p \cap L^1)$$

Since $f_+ \wedge f_- = 0$ and $|Tf| \leq Tf_+ \vee -T(-f_-)$ we finally obtain

$$\|Tf\|^p \leq \|Tf_+\|_p^p + \|T(-f_-)\|_p^p \leq \|f_+\|_p^p + \|f_-\|_p^p = \|f\|_p^p \quad ∎$$

REMARK 2.6: In most of our results T will be order preserving, order continuous, L^1 norm decreasing and L^∞ nonexpansive. With the help of Corollary 2.5 one can then extend the operator T to a mapping with $T(L^p) \subset L^p$. This is described in detail in Krengel and Lin [8],p. 173. We only use the following properties of this extension

(i) $\quad f \leq g \Longrightarrow Tf \leq Tg \quad (f, g \in L^p)$

(ii) $\quad f_n \uparrow f \Longrightarrow Tf_n \uparrow Tf, \; T(-f_n) \downarrow T(-f) \quad (\{f, f_n, n \in \mathbf{N}\} \subset L_+^p)$

(iii) $\quad \|Tf\|_p \leq \|f\|_p \quad (f \in L^p)$

All the subsequent results still hold if one uses another extension method satisfying the above three conditions.

If the above conditions are met then we will use the above L^p extension without any comment.

The next result is a very special case of an interpolation theorem of Browder [2]. Since it follows easily also from 2.5 we include the proof here.

COROLLARY 2.7. *Let T be an order preserving mapping in L^1 which is nonexpansive in L^1 and in L^∞. Then T is also nonexpansive in L^p for any $1 < p < \infty$.*

PROOF: For a fixed $g \in L^1 \cap L^\infty$ the mapping $T_g(f) := T(f+g) - Tg$ is norm decreasing in L^1 and L^∞ nonexpansive and by Corollary 2.5 also L^p–norm decreasing for any $1 < p < \infty$. Thus we have

$$\|Tf - Tg\|_p = \|T_g(f-g)\|_p \leq \|f - g\|_p \quad (f \in L^1 \cap L^p)$$

Since T is nonexpansive in L^1 this can be extended to hold for all $g \in L^p$. ∎

PROPOSITION 2.8. *Assume that $\mu(E) < \infty$. Let T be order preserving, order continuous, and positively integral preserving on L_+^1. If T is L^p nonexpansive for some $1 < p \leq \infty$, then*

(i) $$T(f+t) = Tf + t \quad (f \in L_+^1, \, t \in [0, \infty[)$$

If the above assumptions hold on all of L^1, then

(ii) $$T(f+t) = Tf + t \quad (f \in L^1, \, t \in]-\infty, \infty[)$$

In both cases T is nonexpansive in L_+^p (resp. L^p) for any $1 \leq p \leq \infty$.

PROOF: We may assume $\mu(E) = 1$ and prove (i) and (ii) simultaneously. To this end let $1 < p \leq \infty$ be such that T is L^p nonexpansive on L_+^p (resp. L^p) and let $1 \leq q < \infty$ be such

that $\frac{1}{p}+\frac{1}{q}=1$. Let further $f \in L_+^p$ (resp. $f \in L^p$) and $t \in [0,\infty[$ (resp. $t \in]-\infty,\infty[$) be given. Because T is integral preserving on L_+^1 (resp. L^1) we may use Hölder's inequality to obtain

$$\begin{aligned} |t| &= \left|\int (f+t) - f\,d\mu\right| \\ &= \left|\int T(f+t) - Tf\,d\mu\right| \le \int |T(f+t) - Tf|\,d\mu \le \|T(f+t) - Tf\|_p \|1\|_q \le |t| \end{aligned}$$

Thus we must have equality. This, however, can only happen if $c := T(f+t) - Tf$ is a.e. constant and $|c| = |t|$. Since T is order preserving we must have $c = t$.

Let now $f, g \in L_+^\infty$ (resp. $f, g \in L^\infty$) and $t := \|f-g\|_\infty$. Then $g \le f + t$ and (i) (resp. (ii)) implies $Tg \le T(f+t) = Tf + t$. Interchanging f and g we see that T is L^∞ nonexpansive. Integral preserving implies L^1 nonexpansiveness by [8],Lemma 2.2. By Corollary 2.7 T is L^p nonexpansive for any $1 \le p \le \infty$. ∎

REMARK 2.9: The above proof with $f = 0$ shows that if we assume T to be L^p norm decreasing for some $1 < p \le \infty$ (instead of L^p nonexpansive), then T is L^∞ norm decreasing, and, by Corollary 2.5, it is L^p norm decreasing for any $1 \le p \le \infty$.

PROPOSITION 2.10. *Assume that $\mu(E) < \infty$.*

Let T be an order preserving and positively integral preserving mapping on $L_+^1(\mu)$ such that

$$\text{(i)} \qquad T(f+t) = Tf + t \quad (f \in L_+^1, t \in [0,\infty[)$$

Then there exists a mapping $\overline{T}$ on L^1 with $\overline{T}_{|L_+^1} = T$ which is order preserving, nonexpansive in L^1 and L^∞, integral preserving and which satisfies (i) for every $f \in L^1, t \in]-\infty,+\infty[$.

PROOF: If $f \in L^1$ is bounded below then we define $\overline{T}f = T(f+t) - t$ where $t \in [0,\infty[$ is chosen large enough such that $f + t \ge 0$. By (i) this definition does not depend on the special choice of t. By Proposition 2 of Crandall and Tartar [3] $\overline{T}$ is nonexpansive in L^∞. By the fact that T is positively integral preserving and Lemma 2.2 of [8] T is nonexpansive in L_+^1. Clearly this property then holds also for $\overline{T}$ on the set of all lower bounded elements of L^1. Since this set is dense in L^1 we may extend $\overline{T}$ to all of L^1 such that $\overline{T}$ is still L^1 nonexpansive. By continuity also property 2.8(ii), which holds trivially for functions which are bounded from below, extends to all of L^1. ∎

3. THE S_n ITERATION

For any mapping T on L^1 and $f \in L^1(\mu)$ we define the "nonlinear sums" $S_0 f = f$, $S_{n+1} f = f + T(S_n f)$, the averages $A_n f = S_n f/(n+1)$ and the maximal function

$$M(T)f = Mf := \sup_{0 \le n < \infty} \frac{1}{n+1} S_n f = \sup_{n \in \mathbf{N}} A_n f$$

If T has an extension to other L^p spaces we define the above operators also for these spaces.

PROPOSITION 3.1 (MAXIMAL ERGODIC LEMMA). *Let T be order preserving on L^1 and L^1 norm decreasing on L^1_+. Let further $f \in L^1(\mu)$ and $h \geq 0$ be measurable such that*

$$Tf - th \leq T(f - th)_+ \quad (t \in [0, \infty[)$$

Then we have

$$\int_{\{Mf>h\}} h\, d\mu \leq \int_{\{Mf>h\}} f\, d\mu$$

PROOF: Defining

$$m_n := \sup_{0 \leq k \leq n} (S_k f - (k+1)h)_+ \quad (n \in \mathbf{N})$$

we clearly have

$$(1) \qquad \{Mf > h\} = \bigcup_{n=1}^{\infty} \{m_n > 0\}$$

Setting $S_{-1} = 0$ we see inductively that

$$S_k f - (k+1)h = f - h + T(S_{k-1} f) - kh \leq f - h + T((S_{k-1} f) - kh)_+ \quad (k \geq 0)$$

Because T is order preserving this implies

$$S_k f - (k+1)h \leq f - h + \sup_{-1 \leq j \leq n-1} T(S_j f - (j+1)h)_+ \leq f - h + Tm_{n-1} \quad (0 \leq k \leq n)$$

Taking the maximum on the left side we obtain, since $m_{n-1} \leq m_n$

$$m_n = I_{\{m_n>0\}} \sup_{0 \leq k \leq n} (S_k f - (k+1)h) \leq I_{\{m_n>0\}}(f - h) + Tm_n \quad (n \in \mathbf{N})$$

Putting h and Tm_n on the left side, integrating this inequality and using the fact that T is $L^1(\mu)$–norm decreasing we get

$$\int_{\{m_n>0\}} h\, d\mu \leq \int_{\{m_n>0\}} h\, d\mu + \|m_n\|_1 - \|Tm_n\|_1 \leq \int_{\{m_n>0\}} f\, d\mu$$

Combining this with (1) the assertion follows from the monotone convergence theorem. ∎

REMARK: $h = 0$ satisfies the above assumptions. The general formulation is needed in the sequel, because T is nonlinear.

EXAMPLE 3.2: In this example μ is the counting measure on the two point space $E = \{0, 1\}$. For simplicity we identify $L^p(\mu)$ with $\mathbf{R}^2$ and define

$$T(x, y) := (x \vee y, x \wedge y) \quad ((x, y) \in \mathbf{R}^2)$$

Since $T = T^2$ we have

$$\lim_{n \to \infty} \frac{1}{n+1} \sum_{i=0}^{n} T^i(x, y) = (x \vee y, x \wedge y) \quad ((x, y) \in \mathbf{R}^2)$$

On the other hand one can easily show by induction that

$$\begin{matrix} S^{2n}(x,y) = ((n+1)x + ny, nx + (n+1)y) \\ S^{2n+1}(x,y) = ((n+1)(x+y), (n+1)(x+y)) \end{matrix} \quad (x \le y,\, n \ge 0)$$

Hence if $x < y$ then the two averages have different limits.

Taking $f = (-2,1)$ this example also shows that in general $|Tf| \le T|f|$ is false. Taking $f = (-1,1)$ we see that $Tf = Tf_+ - Tf_-$ is also false in general.

The next example shows that the limit of $\frac{1}{n+1} S_n f$ need not be invariant.

EXAMPLE 3.3: Let (E,μ) be as in example 3.2. Again we identify $L^p(\mu)$ with $\mathbf{R}^2$. It is easily seen that the operator

$$T(x,y) = ((x-1)_+, (y-1)_+) + (y \wedge 1, x \wedge 1)$$

is order preserving, nonexpansive in $L^1(\mu)$, disjointly additive and $T(c,c) = (c,c)$ for any $c \in \mathbf{R}$. Since $S_1(1,0) = (1,1)$ we have $S_n(1,0) = (n,1)$ and therefore $\frac{1}{n+1} S_n(1,0)$ converges to $(1,0)$ which is obviously not invariant. More generally one can easily show the following "bumerang" property

$$\lim_{n\to\infty} \frac{1}{n+1} S_n(x,y) = (x,y) \quad (\,(x,y) \ge (0,0)\,).$$

The above example is much related to a similar one due to Lin and Sine [**9**]. There a mapping T on $L^1([0,1])$ and a function $f \in L^1_+$ is constructed for which $\lim_{n\to\infty} \frac{1}{n+1} \sum_{i=0}^{n} T^i f$ exists but is not invariant. It can be shown that for the same function also $A_n f$ converges to a non–invariant function. Note that in the present much simpler example the Cesarò averages do always converge to an invariant function.

Unlike the first example, the mapping of the last one is not positively homogeneous.

LEMMA 3.4. *Let $1 \le p \le \infty$. If T is also positively homogeneous and nonexpansive in $L^p(\mu)$ then for any $f \in L^p(\mu)$ the limit of $A_n f$ is invariant if it exists.*

PROOF: T is continuous, thus if $A_n f \to h$, then $Th = h$ by the relation

$$A_{n+1} f = \frac{1}{n+2} S_{n+1} f = \frac{1}{n+2} f + \frac{n+1}{n+2} T(A_n f) \quad \blacksquare$$

4. THE DOMINATED ERGODIC THEOREM

We denote by $L \log L$ the space of all measurable functions satisfying

$$\int |f(x)| (\log |f(x)|)_+ \, d\mu < \infty.$$

This space will be of interest to us only if $\mu(E) < \infty$. In this case we have $L \log L \supset L^p$ for any $p > 1$.

Since $|S_n f| \le S_n |f|$ need not hold in general (see Example 3.2), we define

$$\overline{M} f = \sup_{n \in \mathbf{N}} |A_n f|$$

whenever this makes sense.

PROPOSITION 4.1 (DOMINATED ERGODIC THEOREM). *For any order preserving, order continuous, L^1 norm decreasing and L^∞ nonexpansive mapping T on $L^1(\mu)$ we have*

$$\text{(i)} \qquad \mu(\{Mf > t\}) \leq \frac{1}{t}\int_{\{Mf>t\}} f\,d\mu \quad (f \in L^1,\ t > 0)$$

$$\text{(ii)} \qquad \mu(\{\overline{M}f > t\}) \leq \frac{1}{t}\int_{\{\overline{M}f>t\}} |f|\,d\mu \quad (f \in L^1,\ t > 0)$$

$$\text{(iii)} \qquad \|\overline{M}f\|_p \leq \frac{p}{p-1}\|f\|_p \quad (f \in L^p_+,\ 1 < p < \infty)$$

If moreover $\mu(E) < \infty$ then

$$\text{(iv)} \qquad \|\overline{M}f\|_1 \leq \frac{e}{e-1}(\mu(E) + \int f(x)(\log f(x))_+\,d\mu) \quad (f \in L\log L_+)$$

PROOF: Using 2.1(i), property (i) follows from Proposition 3.1 if we take $h = t$.

To prove (ii) define $\tilde{T}f = -T(-f)$. $\tilde{T}$ clearly satisfies all the assumptions of the proposition. Its nonlinear sums $\tilde{S}_0 f = f$, $\tilde{S}_{n+1}f = f + \tilde{T}(\tilde{S}_n f)$ satisfy $\tilde{S}_n f = -S_n(-f)$ for $n \geq 0$ as is easily proved by induction. Hence, denoting $\tilde{M} = M(\tilde{T})$, (i) yields

$$\begin{aligned}\mu(\{\overline{M}f > t\}) &\leq \mu(\{Mf > t\}) + \mu(\{\tilde{M}(-f) > t\}) \\ &\leq \frac{1}{t}\int_{\{Mf>t\}} f_+\,d\mu + \frac{1}{t}\int_{\{\tilde{M}(-f)>t\}} f_-\,d\mu \leq \frac{1}{t}\int_{\{\overline{M}f>t\}} |f|\,d\mu\end{aligned}$$

By (ii), $|f|$ and $\overline{M}f$ are in maximal type relation (see Krengel [**5**],p. 52). Using Lemma 1.6.2 of [**5**] it follows, just as in the proof of [**5**] Theorem 1.6.3, that (iii) and (iv) hold at least for $f \in L^\infty$. The general case follows by suitable approximation and order continuity 2.6(ii). ∎

5. POINTWISE ERGODIC THEOREMS

In this section T will always be *order preserving, order continuous, L^1 norm decreasing, L^∞ nonexpansive and positively homogeneous.*

Then T is well defined on all L^p spaces and we put

$$L^p_I := \{h \in L^p : Th = h\}, \quad L^p_{I+} = \{h \in L^p_I : h \geq 0\} \quad (1 \leq p < \infty)$$

THEOREM 5.1. *Let $1 < p < \infty$ be given. If*

$$\text{(i)} \qquad L^p_{I+} \cap L^1 = \{0\}$$

then for any $f \in L^p_+$ we have

$$\text{(ii)} \qquad \lim_{n\to\infty} A_n f = 0 \quad \mu\text{-a.e.}$$

$$\text{(iii)} \qquad \lim_{n\to\infty} \|A_n f\|_p = 0$$

(ii) and (iii) hold for all $f \in L^p$, if

$$\text{(iv)} \qquad L^p_I \cap L^1 = \{0\}.$$

PROOF: Let $f \in L^p$. By the dominated ergodic theorem we have $|A_n f| \leq \overline{M}f \in L^p$, whence $g_1 = \limsup_{n\to\infty} A_n f$ and $g_2 = \liminf_{n\to\infty} A_n f$ are also in L^p. Since by positive homogeneity

$$\frac{n+2}{n+1} A_{n+1} f = \frac{1}{n+1} f + T(A_n f)$$

we have

$$g_1 = \limsup_{n\to\infty} T(A_n f) \leq T(\limsup_{n\to\infty} A_n f) = T g_1$$

and similarly $g_2 \geq T g_2$. Hence $T^k g_1$ increases to a function $\overline{g} \leq \overline{M}f$. Since T is L^p norm decreasing by Corollary 2.5. $\overline{g}$ belongs again to L^p. Of course we have also $\overline{g} \in L^p_I$. Similarly, $T^k g_2$ decreases to an invariant function $\underline{g}$ in L^p.

Thus if either (i) holds and $f \geq 0$ or (iv) holds, then (ii) follows from $0 = \underline{g} \leq g_1 \leq g_2 \leq \overline{g} = 0$. By the dominated convergence theorem (iii) holds too. ∎

REMARKS: (1) For $f \in L^p_+$ the proof that $Mf \in L^p$ (Proposition 4.1) requires only T to be L^1 norm decreasing in L^1_+, hence the conclusions of Theorem 5.1 still hold for $f \in L^p_+$ (under hypothesis (i)) if T is L^1 norm decreasing in L^1_+ (with the same proof). This will be used for some examples in sect. 6.

(2) If there exists an invariant function $0 \neq g \in L^p$ then $A_n g = g$, whence convergence to 0 is impossible. Thus (i) or (iv) are natural in infinite measure spaces.

A measurable set $A \subset E$ is called *invariant* if $\mu(A) < \infty$ and $TI_A = I_A$. The set of all invariant subsets will be denoted by Σ_I.

To verify condition 5.1(i) the following lemma may be helpful.

LEMMA 5.2. *Assume that T is also positively integral preserving. Let $1 \leq p < \infty$ and $h \in L^p_I$. Then for $t > 0$ we have $\{h > t\} \in \Sigma_I$.*

Hence condition 5.1(i) holds if and only if $\Sigma_I = \{A \subset E$ measurable, $\mu(A) = 0\}$.

PROOF: Let $h \in L^p_I$ and $t > 0$ be given. By 2.1(i), which extends to L^p, we have

(1) $$(h-t)_+ = (Th-t)_+ \leq T(h-t)_+ \quad (t > 0)$$

On the other hand, since T is L^1–norm decreasing and since $(f-t)_+ \in L^1$ for any $f \in L^p$ and $t > 0$, we have $\|T(h-t)_+\|_1 \leq \|(h-t)_+\|_1$. Together with (1) we obtain $(h-t)_+ = T(h-t)_+$. For any $s > 0$ we now have

$$T((h-t)_+ \wedge s) \leq T(h-t)_+ \wedge Ts \leq (h-t)_+ \wedge s.$$

Since T is positively integral preserving, we have

$$\int T((h-t)_+ \wedge s)\, d\mu = \int (h-t)_+ \wedge s\, d\mu$$

Hence we must have $T((h-t)_+ \wedge s) = (h-t)_+ \wedge s$. Since obviously $n((h-t)_+ \wedge \frac{1}{n}) \uparrow I_{\{h>t\}}$, order continuity and positive homogeneity imply

$$TI_{\{h>t\}} = T(\sup_{n\in\mathbf{N}} n((h-t)_+ \wedge \tfrac{1}{n})) = \sup_{n\in\mathbf{N}} T(n((h-t)_+ \wedge \tfrac{1}{n})) = \sup_{n\in\mathbf{N}} n((h-t)_+ \wedge \tfrac{1}{n}) = I_{\{h>t\}} \quad ∎$$

THEOREM 5.3. *Let $\mu(E) < \infty$ and T be integral preserving. If Σ_I is trivial, then*

(i)
$$\lim_{n\to\infty} A_n f = \mu(E)^{-1} \int f \, d\mu \quad \text{a.e.} \quad (f \in L \log L)$$

(ii)
$$\lim_{n\to\infty} \|A_n f - \mu(E)^{-1} \int f \, d\mu\|_p = 0 \quad (f \in L^p, 1 \le p < \infty)$$

PROOF: (i) Let $f \in L \log L$. By Propsition 4.1, $h := \sup_{n\in\mathbf{N}} |A_n f| \in L^1$. Define $g := \limsup_{n\to\infty} A_n f$. then $|g| \le h \in L^1$. Let $\ell_n := \sup_{k\ge n} A_k f$. Then $|\ell_n| \le h \in L^1$ and $\ell_n \downarrow g$. Now

$$T\ell_n \ge \sup_{k\ge n} TA_k f = \sup_{k\ge n} \left(\frac{k+2}{k+1} A_{k+1} f + \frac{f}{k+1} \right) \ge \sup_{k\ge n} A_{k+1} f - \frac{f}{n+1} = \ell_{n+1} - \frac{f}{n+1}$$

By continuity in L^1, $Tg = \lim_{n\to\infty} T\ell_n \ge g$. Since T is integral preserving and $g \in L^1$, we obtain $Tg = g$. By Lemma 5.2, $\{g > t\} \in \Sigma_I$ for any $t > 0$. By Proposition 2.8(ii) we have $T(f + r) = Tf + r$ for $-\infty < r < \infty$, and by induction $S_n(f + r) = S_n f + (n + 1)r$. Hence $A_n(f + r) = A_n f + r$ and $\limsup_{n\to\infty} A_n(f + r) = g + r$ for any r. Applying the above to $f + r$ we have $\{g + r > t\} \in \Sigma_I$ for $t > 0$ and $r \in \mathbf{R}$. Hence $\{g > t\} \in \Sigma_I$ for any $t \in \mathbf{R}$.

We now assume $\int f \, d\mu = 0$, and show $g \le 0$. To this end, assume $\mu\{g > t\} > 0$ for some $t > 0$. Then, since Σ_I is trivial, $\{g > t\} = E$. Let $f_t = f - t$. Then $g > t$ a.e. implies $Mf_t = (Mf) - t > 0$ a.e., and by the maximal ergodic lemma 3.1 (with $h = 0$), we get the contradiction

$$0 \le \int_{\{Mf_t > 0\}} f_t \, d\mu = \int f_t \, d\mu = \int f \, d\mu - t\mu(E) = -t\mu(E) < 0$$

Thus $g \le 0$. To show $\liminf_{n\to\infty} A_n f \ge 0$ we use $\tilde{T} f = -T(-f)$ (as in the proof of Proposition 4.1). Since $\tilde{A}_n f = -A_n(-f)$, we apply our lim sup result to $\tilde{T}$ to obtain

$$-\liminf_{n\to\infty} A_n f = \limsup_{n\to\infty} -A_n f = \limsup_{n\to\infty} \tilde{A}_n(-f) \le 0.$$

If $f \in L \log L$ is arbitrary, let $r = \mu(E)^{-1} \int f \, d\mu$, then we have

$$\lim_{n\to\infty} A_n f = r + \lim_{n\to\infty} A_n(f - r) = r + 0 \quad \text{a.e.}$$

(ii) follows from (i) by the bounded convergence theorem if $f \in L^p$, $1 < p < \infty$ and in L^1 for bounded functions. Since T is also nonexpansive in L^1, standard approximations yield (ii) also in L^1. ∎

REMARK 5.4: Assume that T satisfies all the assumptions of Theorem 5.3 except that instead of L^∞ nonexpansiveness we assume only that T is L^∞ norm decreasing. Let further $f \in L^\infty$ with $\int f \, d\mu < 0$. Then $g := \limsup_{n\to\infty} A_n f$ is trivially in L^∞, because T is L^∞ norm decreasing. Thus the dominated ergodic Theorem, which requires L^∞ nonexpansiveness, is no more needed. Furthermore the proof of Lemma 5.2 (with $t = 0$) shows that $\{g > 0\} \in \Sigma_I$. Applying the maximal ergodic theorem as above, we obtain $g \le 0$.

Let now $f \in L^\infty$ with $\int f\,d\mu = 0$ then, for any $\varepsilon > 0$ we may apply the above argument to $f - \varepsilon$ and, replacing T by $\tilde{T}$, to $-f - \varepsilon$. Using L^1 nonexpansiveness and making ε smaller and smaller, this implies convergence in measure of $A_n f$ to 0. Because of uniform boundedness we have then also norm convergence in L^p for $(A_n f)$.

EXAMPLE 5.5: Define T on $\mathbf{R}^2$ by $T(x,y) = (\frac{1}{3}(x \vee y) + x + y, \frac{1}{3}(x \wedge y) + x + y)$.

EXAMPLE 5.6: Let E be the unit interval with Lebesgue measure. Let T_1 be defined by

$$T_1 f(x) = \begin{cases} f(x) \vee f(x+1/2) & (0 \le x < 1/2) \\ f(x) \wedge f(x+1/2) & (1/2 \le x < 1) \end{cases}$$

(T_1 is like T in Example 3.2, but on a continuous space). Let now $0 < \alpha < 1/2$ be fixed and define

$$T_2 f(x) = \int_{-\alpha}^{\alpha} f((x+y) \mod 1)\,dy$$

Then T_2 is a linear order preserving contraction such that $T_2 f$ is continuous for any $f \in L^\infty$. We now define $T := T_2 \circ T_1$. Since $TI_A = T_2(T_1 I_A)$ is continuous, I_A must be continuous for any $A \in \Sigma_I$. Thus Σ_I is trivial. Clearly T satisfies all the other assumptions of the above theorem.

The assumption $\Sigma_I = \{\emptyset, E\}$ is the "ergodic" case. For the nonergodic case we need (as in the linear case) a study of the invariant functions.

T is said to be ***decomposable by invariant sets*** (shortly (DIS)) if for any $A \in \Sigma_I$ we have

$$I_A f = 0 \Longrightarrow I_A T f = 0 \quad (f \in L^1_+).$$

$L^1_+(A)$ is always T–invariant for any $A \in \Sigma_I$. (DIS) requires $L^1_+(E \setminus A)$ also to be T–invariant.

In the linear case, when T is a contraction possessing an equivalent finite invariant measure, this is no restriction at all. However, as the invariant set $\{0\}$ in example 3.2 shows, this condition is not necessarily fulfilled in the nonlinear case. In fact, all speed limit operators on a *finite* interval (see Krengel [7]) violate this condition. On the other hand it is usually very easily decided whether (DIS) holds or not.

THEOREM 5.7. *Assume that $\mu(E) < \infty$.*

Assume that T is order preserving, order continuous, L^∞ nonexpansive, integral preserving, positively homogeneous and that (DIS) holds. Then we have

(i) Σ_I *is a σ-algebra on E*

(ii) $L^p_I = L^p(\mu_{|\Sigma_I}) \quad (1 \le p < \infty)$

(iii) $T(hf) = hTf \quad (h \in L^1_{I+}, f \in L^1, (fh) \in L^1)$

(iv) $T(h+f) = h + Tf \quad (h \in L^1_I, f \in L^1)$

For any $f \in L^1$ we denote by $P_I f$ the Radon–Nikodym derivative of the signed measure $(f\mu)_{|\Sigma_I}$ with respect to $\mu_{|\Sigma_I}$. (If μ is a probability measure, then $P_I f$ is just the conditional expection with respect to the σ-algebra Σ_I.)

(v) *For $1 \le p < \infty$ P_I is a positive, linear contractive projection of L^p onto L^p_I.*

(vi) $TP_I = P_I = P_I T$

(vii) $\lim_{n\to\infty} A_n f = P_I f \quad \mu\text{-a.e.} \quad (f \in L \log L)$

(viii)

$$\lim_{n\to\infty} \|A_n f - P_I f\|_p = 0 \quad (f \in L^p, 1 \le p < \infty)$$

The proof of Theorem 5.7 is rather long and since (DIS) excludes the most interesting examples we omit it. Nevertheless there exist many examples :

EXAMPLE 5.8: Let T be a mapping on L^1 which satisfies all the assumptions of Theorem 5.7 except (DIS). Then for any $0 < t < 1$ the mapping $S_t f := tTf - (1-t)T(-f)$ satisfies also (DIS).

To see this, let $A \in \Sigma_I(S_t)$ be given. Since tTI_A and $-(1-t)T(-I_A)$ are both nonnegative we have $0 \le TI_A \le I_A$ and $0 \le -T(-I_A) = 1 - TI_{E\setminus A} \le I_A$. Since T is also integral preserving we must have equality in the last two inequalities. Thus $A, E\setminus A \in \Sigma_I$ and (DIS) follows.

Theorem 5.1 and Theorem 5.7 can be combined to yield the following :

THEOREM 5.9. *Assume that T is order preserving, order continuous, L^1 norm decreasing, L^∞ nonexpansive, positively integral preserving, positively homogeneous and that (DIS) holds. Let E_I be defined by $I_{E_I} = \sup_{A\in\Sigma_I} I_A$ (not the pointwise sup but the sup in L^1). Let further $\overline{\Sigma}_I$ be the σ-algebra on E_I generated by Σ_I. For any $f \in L^1$ we denote by $P_I f$ be the function which is on E_I the Radon–Nikodym derivative of the signed measure $(f\mu)_{|\overline{\Sigma}_I}$ with respect to $\mu_{|\overline{\Sigma}_I}$ and is 0 elsewhere. Then we have*

(i) $$\lim_{n\to\infty} A_n f = P_I f \quad \mu\text{-a.e.} \quad (f \in L^p_+, 1 < p < \infty)$$

(ii) $$\lim_{n\to\infty} \|A_n f - P_I f\|_p = 0 \quad (f \in L^p_+, 1 < p < \infty)$$

6. GENERALIZED MEASURE PRESERVING TRANSFORMATIONS

Let $\mathcal{B}$ be the σ-algebra of measurable sets on E. Following Krengel [7] we say that a mapping $\phi : \mathcal{B} \longrightarrow \mathcal{B}$ is a *generalized measure preserving transformation* (shortly *gmpt*) if

(i) $$A \subset B \Longrightarrow \phi(A) \subset \phi(B) \quad (A, B \in \mathcal{B})$$

(ii) $$\mu(\phi(A)) = \mu(A) \quad (A \in \mathcal{B})$$

Clearly, if $\tau : E \longrightarrow E$ is a measure preserving transformation, then $\phi_\tau(A) := \tau^{-1}(A)$ is a generalized measure preserving transformation, satisfying $\phi(E) = E$.

On the space $\mathcal{F}$ of a.e. finite, measurable functions we define T by

$$Tf(x) = I_{\phi(E)}(x) \sup\{t \in \mathbf{R} : x \in \phi(\{f > t\}\}$$

$T = T_\phi$ is said to be *associated* with the generalized measure preserving transformation ϕ and is shortly called a *gmpt* mapping. It is characterized by the property $\{Tf > t\} = \phi\{f > t\}$.

Even if T_ϕ is linear, ϕ need not be induced by a measure preserving transformation. $\phi(A) = 1 + A$ on $\mathbf{N}$ is such a gmpt.

Krengel [7] studied T only on nonnegative functions and obtained the following result :

PROPOSITION 6.1. *If T is a gmpt mapping, then T maps L^1_+ into itsself, is order preserving, positively integral preserving and L^∞ nonexpansive on L^1_+.*

We list some further elementary properties of T_ϕ and ϕ.

PROPOSITION 6.2. *Let $\mathcal{F}_{\text{fin}} = \{f \in \mathcal{F} : f \geq 0,\ \mu\{f > t\} < \infty \forall t > 0\}$ and $T = T_\phi$ be associated with a gmpt ϕ. Then we have*

(i) $B_n \uparrow B \Longrightarrow \phi(B_n) \uparrow \phi(B) \quad (\{B, B_n, n \in \mathbf{N}\} \subset \mathcal{B},\ \mu(B) < \infty)$

(ii) $B_n \downarrow B \Longrightarrow \phi(B_n) \downarrow \phi(B) \quad (\{B, B_n, n \in \mathbf{N}\} \subset \mathcal{B},\ \mu(B_1) < \infty)$

(iii) $Tf(x) = I_{\phi(E)}(x) \sup\{t \in \mathbf{R} : x \in \phi(\{f \geq t\}\} \quad (f \in \mathcal{F}, f \in \mathcal{F}_{\text{fin}})$

(iv) $f_n \uparrow f \Longrightarrow Tf_n \uparrow Tf \quad (\{f, f_n, n \in \mathbf{N}\} \subset \mathcal{F}_{\text{fin}})$

(v) $f_n \downarrow f \Longrightarrow Tf_n \downarrow Tf \quad (\{f, f_n, n \in \mathbf{N}\} \subset \mathcal{F}_{\text{fin}})$

(vi) $T(\mathcal{F}) \subset \mathcal{F}$

(vii) $T(\alpha f) = \alpha Tf \quad (\alpha \geq 0, f \in \mathcal{F})$

(viii) $T(f + \alpha) = I_{\phi(E)}((Tf) + \alpha) \quad (f \in \mathcal{F}, \alpha \in \mathbf{R})$

(ix) $\mu(E) < \infty \Longrightarrow \phi(E) = E$

Using some of the above preoperties we obtain the following

PROPOSITION 6.3. *Let T be a gmpt mapping. If $\mu(E) < \infty$ then T is order preserving, nonexpansive in L^p for any $1 \leq p \leq \infty$, integral preserving, and positively homogeneous.*

REMARKS: (a) If $\mu(E) < \infty$ then our definition of T_ϕ on all of L^1 is just the extension provided by Proposition 2.10 of the original T_ϕ of Krengel, defined only on L^1_+.
(b) In Example 3.2 T is associated with the gmpt ϕ defined by

$$\phi(\{0\}) = \phi(\{1\}) = \{0\}, \quad \phi(\{0,1\}) = \{0,1\}$$

If $\mu(E) = \infty$ then the situation is less pleasing. Even $T_\phi(L^1) \subset L^1$ may fail. For this purpose we need the following

LEMMA 6.4. *Let T be a gmpt mapping. Then we have $(Tf)_+ = T(f_+)$ and $-(Tf)_- = T(-f_-)$ for any $f \in \mathcal{F}$.*

PROOF: Let $x \in \phi(E)$ be given. If there exists $s \geq 0$ with $x \in \phi\{f > s\}$ then $x \in \phi\{f_+ > s\}$ $= \phi\{f > s\}$ and $x \in \phi\{f > s\} \subset \phi\{-f_- > t\}$ for any $t < 0$. Hence $Tf(x) = Tf_+(x)$ and $T(-f_-)(x) = 0$. On the other hand, if $x \notin \phi\{f > 0\}$, then $x \notin \phi\{f_+ > 0\}$ and $\{f > t\} =$ $\{-f_- > t\}$ for any $t < 0$. Hence $Tf_+(x) = 0$ and $Tf(x) = T(-f_-)(x)$. ∎

PROPOSITION 6.5. *Let ϕ be a gmpt. Then $T = T_\phi$ is L^1 norm decreasing on L^1 (in particular, $T(L^1) \subset L^1$) if and only if*

(i) $$\mu(E \setminus \phi(E \setminus A)) \leq \mu(A) \quad (A \in \mathcal{B})$$

PROOF: Since $T(-1_A) = -1_{E \setminus \phi(E \setminus A)}$ condition (i) is clearly necessary.

In view of Proposition 6.1 and Lemma 6.4 it suffices to show the assertion for nonpositive functions $f \in L^1$. Since $\{Tf \leq t\} = E \setminus \{Tf > t\} = E \setminus \phi\{f > t\} = E \setminus \phi(E \setminus \{f \leq t\})$ this follows from

$$\|f\|_1 = \int_{-\infty}^{0} \mu\{Tf \leq t\}\, dt = \int_{-\infty}^{0} \mu(E \setminus \phi(E \setminus \{f \leq t\}))\, dt \leq \int_{-\infty}^{0} \mu(\phi(\{f \leq t\}))\, dt = \|f\|_1 \quad ∎$$

REMARKS: (a) If we have equality in (i) then the above proof shows that T is even integral preserving on L^1. In particular, such a mapping is L^1 nonexpansive on the whole L^1 and therefore, by interpolation, also L^p nonexpansive.
(b) The speed limit operators of [8] satisfy (i). This can be easily seen if A is a finite union of disjoint intervals and for general A this follows by a standard approximation argument. In an area where the speed function is nonconstant, we have $\mu(E \setminus \phi(E \setminus A)) < \mu(A)$ and therefore T is not integral preserving on negative functions. Nevertheless, it can be shown that speed limit operators are not only L^1 norm decreasing (as follows from the above proposition), but even L^1 nonexpansive on L^1.

We are now going to apply our convergence theorems to generalized measure preserving transformations. It turns out, that in this special case we can prove a.e. convergence for all functions in L^1_+.

THEOREM 6.6. *Let T be a gmpt mapping. If $\Sigma_I = \{A \in \mathcal{B} : \mu(A) = 0\}$, then*

$$\text{(i)} \qquad \lim_{n\to\infty} A_n f = 0 \text{ a.e.} \quad (1 \le p < \infty, f \in L^p_+)$$

$$\text{(ii)} \qquad \lim_{n\to\infty} \|A_n f\|_p = 0 \quad (1 < p < \infty, f \in L^p_+)$$

PROOF: By Lemma 5.2 the assumptions of Theorem 5.1 are fulfilled and both assertions follow for $1 < p < \infty$.

For $p = 1$, note that, for $f \in L^1_+$, Mf is a.e. finite by the dominated ergodic theorem 4.1. In particular, $g = \limsup_{n\to\infty} A_n f \ge 0$ is a.e. finite. Since T is a gmpt mapping $Tg \ge 0$ is a.e. finite too. Let $\ell_n = \sup_{k\ge n} A_k f$. Then $\ell_n \downarrow g$ and, by 4.1(i) and 6.2(v), also $T\ell_n \downarrow Tg$ a.e.. Together with

$$T\ell_n \ge \sup_{k\ge n} \frac{TS_k f}{k+1} = \sup_{k\ge n} \frac{S_{k+1} f - f}{k+1} \ge \sup_{k\ge n} \frac{S_{k+1} f}{k+1} - \frac{f}{n+1} = \ell_{n+1} - \frac{f}{n+1}$$

this implies

$$Tg = \lim_{n\to\infty} T\ell_n \ge \limsup_{n\to\infty} \left[\ell_{n+1} - \frac{f}{n+1}\right] = g$$

For any $t > 0$ $\{Tg > t\} = \phi\{g > t\}$ has finite measure, by the maximal ergodic lemma 3.1, and $\mu\{Tg > t\} = \mu\{g > t\} < \infty$. Hence $Tg = g$. In general, however, $Mf \notin L^1$ and so Lemma 5.2 is not directly applicable. On the other hand, for any $0 < s < t < \infty$ we have $g_{st} := \inf((g - s)_+, t) \in L^1$. If we can also show that $Tg_{st} = g_{st}$, then by Lemma 5.2 $g_{st} = 0$ and therefore $g = 0$ which finishes the proof. To see this we define $g_s = (g - s)_+$. Because T is order preserving and because of 6.2(viii) we have $Tg_s \ge g_s$ and since $\mu\{Tg_s > r\} = \mu\{g_s > r\}$ for any $r > 0$ we have $Tg_s = g_s$. Again, since T is order preserving, we have then $Tg_{st} = T(g_s \wedge t) \le g_s \wedge t = g_{st}$. Again, since $\mu\{Tg_{st} > r\} = \mu\{g_{st} > r\}$ for any $r > 0$, we get finally $Tg_{st} = g_{st}$. ∎

Combining Theorem 5.1 and Proposition 6.5 we obtain the following

THEOREM 6.7. *Let T satisfy the assumption of the previous Theorem and also condition 6.5(i). Let further $1 < p < \infty$ such that*

$$Tg = g \Longrightarrow g = 0 \quad (g \in L^p)$$

Then

$$\text{(i)} \qquad \lim_{n\to\infty} A_n f = 0 \text{ a.e.} \quad (f \in L^p)$$

$$\text{(ii)} \qquad \lim_{n\to\infty} \|A_n f\|_p = 0 \quad (f \in L^p).$$

EXAMPLE 6.8: While in the infinite measure case the above assumption on invariant sets is often met, it is not clear, that truely nonlinear gmpts on finite measure spaces satisfying (DIS) exist. The following is an example, due to T. Hill and U. Krengel, of a nonlinear gmpt on a finite measure space, which doesn't possess any nontrivial invariant set at all.

His example is of the form $\phi = \phi_1 \circ \phi_2$ where ϕ_1 is a mixing measure preserving transformation and ϕ_2 is a very simple gmpt, which even satisfies $\phi_2 \circ \phi_2 = \phi_2$. The details are as follows.

τ is the right shift on $E := \{0,1\}^{\mathbf{Z}}$, i.e. $\tau((x_n)_{n\in\mathbf{Z}}) = (x_{n-1})_{n\in\mathbf{Z}}$. The measure μ on E is just the infinite product of the measure on $\{0,1\}$, which assign mass $1/2$ to each of the two points. Let ϕ_1 be defined by $\phi_1(A) = \tau^{-1}(A)$. It is well known that τ is mixing i.e., for any two measurable sets A, B we have

$$\lim_{n\to\infty} \mu(\phi_1^n(A) \cap B) = \mu(A)\mu(B).$$

Let X_i the projection on the i-th coordinate. Let p be the mapping on E such that $X_0 \circ p = 1$ and $X_n \circ p = X_n$ for any $n \neq 0$. Then for any measurable $A \subset E$ we define

$$\phi_2(A) = p(A) \cup (p^{-1}(A) \cap \{X_0 = 0\} \cap A)$$

For any $n \in \mathbf{Z}$ let σ_n be the σ-algebra generated by $\{X_i : -\infty < i \leq n\}$. Then we have

$$\phi_1(\sigma_n) = \sigma_{n-1}, \qquad \phi_2(\sigma_n) \subset \sigma_n \qquad (-\infty < n < \infty)$$
$$\phi_2(A) = A \quad (-\infty < n < 0,\ A \in \sigma_n)$$

Thus $\phi(\sigma_n) \subset \sigma_{n-1}$ and for $n > 0$ $\phi^n(\sigma_{n-1}) \subset \sigma_{-1}$, whence

$$\phi^m(A) = \phi_1^{m-n} \circ \phi^n(A) \quad (0 < n < m, A \in \sigma_{n-1})$$

Since $\bigcup \sigma_n$ is dense in the entire σ-algebra of E, we can find for any measurable set A and any $\varepsilon > 0$ an $n \in \mathbf{N}$ such that

$$\mu(\phi^m(A) \setminus \phi_1^{m-n} \circ \phi^n(A)) + \mu(\phi_1^{m-n} \circ \phi^n(A) \setminus \phi^m(A)) \leq \varepsilon$$

for any $m \geq n > 0$. Since τ is mixing, this implies

$$\lim_{m\to\infty} \mu(\phi^m(A) \cap B) = \mu(A)\mu(B).$$

for any two measurable sets $A, B \subset E$. In particular, $\phi(A) = A$ implies, that either $\mu(A) = 0$ or $\mu(A) = 1$.

For the averages $\frac{1}{n+1}\sum_{i=0}^{n} T^i f$, when T is nonexpansive, there exist many results about *weak* convergence. Norm convergence could only be shown under the additional condition that T is *odd*, i.e. $T(-f) = -Tf$, and only if T is nonexpansive in a Hilbert space (see Baillon [**1**]). In Wittmann [**10**] Baillon's theorem on norm convergence was generalized to mappings T on arbitrary subsets C of a Hilbert space H which need not be nonexpansive, but satisfy only $\|Tx + Ty\| \leq \|x + y\|$. However, if C is symmetric and convex, this assumption is easily seen to be equivalent with that of Baillon.

Oddness, however, is a very strong condition :

PROPOSITION 6.9. *Let $\mu(E) < \infty$ and T associated with a gmpt ϕ.*
If T is odd then T is a linear isometry in L^p for any $1 \leq p \leq \infty$.

PROOF: For any measurable set A we have

$$TI_{A^c} = 1 + T(I_{A^c} - 1) = 1 + T(-I_A) = 1 - TI_A = I_{\phi(A)^c}$$

and therefore

$$A \cap B = \emptyset \Longrightarrow \phi(A) \cap \phi(B) = \emptyset \quad (A, B \in \mathcal{B})$$

Thus for two disjoint measurable sets A, B we have

$$\begin{aligned} \mu(A) + \mu(B) &= \mu(\phi(A)) + \mu(\phi(B)) = \mu(\phi(A) \cup \phi(B)) \leq \mu(\phi(A \cup B)) \\ &= \mu(A \cup B) = \mu(A) + \mu(B) \end{aligned} \tag{1}$$

If now f, g are measurable step functions then we can find a family $(A_i)_{1 \leq i \leq n}$ of pairwise disjoint sets such that

$$f = \sum_{i=1}^{n} \alpha_i I_{A_i}, \quad g = \sum_{i=1}^{n} \beta_i I_{A_i}$$

where (α_i) and (β_i) are suitable families in $\mathbf{R}$. Using (1) we get

$$T(f+g) = \sum_{i=1}^{n} (\alpha_i + \beta_i) I_{\phi(A_i)} = \sum_{i=1}^{n} \alpha_i I_{\phi(A_i)} + \sum_{i=1}^{n} \beta_i I_{\phi(A_i)} = Tf + Tg$$

Thus T is linear and isometric on step functions. The assertion follows now by continuity. ∎

We now pass to a special class of generalized measure preserving transformations, the class of speed limit operators. These were introduced by Krengel and Lin [8] to give quite natural examples of nonlinear mappings for which the averages $\frac{1}{n+1} \sum_{i=1}^{n} T^i f$ are not norm convergent in general. On the other hand, as we shall see soon, for $A_n f$ we have not only always norm convergence but also a.e. convergence.

From now on μ will be the Lebesgue measure on $E =]-\infty, +\infty[$. For a given decreasing but not necessarily strictly decreasing "speed function" $\psi : \mathbf{R} \longrightarrow]0, \infty[$ Krengel and Lin [8] (see also Krengel [7]) have constructed a semigroup $(\phi_t)_{t \geq 0}$, i.e. $\phi_{s+t} = \phi_s \circ \phi_t$, of generalized measure preserving transformations on E such that

$$\phi_t([a, b[) = [a + \ell(b, t), b + \ell(b, t)[\quad (-\infty < a < b < +\infty,\ t \geq 0) \tag{S}$$

where

$$\ell(x, t) := \sup\{y \geq x : \int_x^y \frac{1}{\psi(s)} ds \leq t\}$$

We refer to [8] for the intuitive background of these transformations. We now have the following

PROPOSITION 6.10. *Let ψ be a as above and $T = T_{\phi_1}$. Then we have*

$$\Sigma_I = \{A \in \mathcal{B} : \mu(A) = 0\}$$

and therefore the assumptions of Theorem 6.6 are fulfilled.

PROOF: We note first that, since ψ has a lower bound > 0 on any finite interval, we have

$$\lim_{t\to\infty} \ell(x,t) = \infty \tag{1}$$

If T_t is the mapping associated with ϕ_t then we have obviously

$$T^n = T_n \tag{2}$$

Now, if there exists $A \in \Sigma_I$ with $\mu(A) > 0$ then we choose $B \subset A \cap [a,b[$ with $\mu(B) > 0$ and $-\infty < a < b < \infty$. Using (S) and (1) we can find an increasing sequence (n_k) in $\mathbf{N}$ with

$$\phi_{n_i}([a,b[) \cap \phi_{n_j}([a,b[) = \emptyset \quad (i,j \in \mathbf{N}, i \neq j)$$

and therefore also

$$\phi_{n_i}(B) \cap \phi_{n_j}(B) = \emptyset \quad (i,j \in \mathbf{N}, i \neq j)$$

Together with (2) we get

$$\mu(\bigcup_{n=1}^{\infty}\{T^n I_B > 0\}) = \sum_{i=1}^{\infty} \mu(\phi_{n_i}(B)) = \infty$$

$$\bigcup_{n=1}^{\infty}\{T^n I_B > 0\} \subset A$$

and therefore we get the contradiction $\mu(A) = \infty$. ∎

Acknowledgement. We are indebted to U. Krengel for allowing us to include the example of an ergodic nonlinear gmpt.

REFERENCES

1. J.-B. Baillon, *Quelques propriétés de convergence asymptotique pour les contractions impaires*, C. R. Acad. Sc. Paris **283** (1976), 587–590.
2. F. E. Browder, *Remarks on nonlinear interpolation in Banach spaces*, J. Funct. Anal. **4** (1969), 390–403.
3. M. G. Crandall, L. Tartar, *Some relations between nonexpansive mappings and order preserving mappings*, Proc. Amer. Math. Soc. **78** (1980), 385–390.
4. A. M. Garsia, "Topics in almost everywhere convergence," Markham Publ. Co., 1970.
5. U. Krengel, "Ergodic Theorems," deGruyter, Berlin – New York, 1985.
6. U. Krengel, *An example concerning the nonlinear pointwise ergodic theorem*, Israel J. Math. **58** (1987), 193–197.
7. U. Krengel, *Generalized measure preserving transformations*, In : A.e. Convergence, Proc. 1988 conference (1989), 215–235, Academic Press.
8. U. Krengel, M. Lin, *Order preserving nonexpansive operators in L^1*, Israel J. Math. **58** (1987), 170–192.
9. M. Lin, R. C. Sine, *On the fixed point set of nonexpansive order preserving maps*, Math. Z. **203** (1990), 227–234.
10. R. Wittmann, *Mean ergodic theorems for nonlinear operators*, Proc. Amer. Math. Soc. **108** (1990), 781–788.

P.O. Box 653, Beer–Sheva 84105, Israel
Fine Hall – Washington Road, Princeton N. J. 08544 – 1000

On the almost sure central limit theorem

by

M. Peligrad* and P. Révész

University of Cincinnati and Technical University of Vienna

*Partially supported by an NSF grant

ISBN 0-12-085520-8

1. Introduction and results

In the recent years, the almost sure central limit theorem was a subject of several papers by Brosamler (1988), Fisher (1989), Lacey and Philipp (1990), Schatte (1988).

In general the authors used different proofs, and they had different motivations to attack this problem derived either from ergodic theory, occupation time for a random walk, to test the random number generators by following only a particular string of numbers or to study the empirical distribution of the partial sums.

All the above papers contain (sometimes using different conditions) the following result:

Theorem A. Let $\{X_n\}$ be an i.i.d. sequence of random variables which are centered with finite second moment. Define $S_n = \sum_{i=1}^{n} X_i$. Then, there is a P-null set N such that for all $\omega \notin N$ and for all Borel set $A \subset R$ with $\lambda(\delta A) = 0$ we have

$$\lim_{n\to\infty} \frac{1}{\log n} \sum_{j=1}^{n} \frac{1}{j} I_A\left(\frac{S_j}{\sqrt{j}}\right) = P(N \in A) \tag{1}$$

where N is the standard normal distribution, λ is the Lebesque measure, δA is the boundary of A and I_A is the indicator function of A.

One method of proving (1) starts by establishing the result for the i.i.d. normal variables and then by transporting the result via an embedding theorem. Lacey and Philipp (1990) use a direct approach based on a blocking argument in connection with a law of large numbers for asymptotically orthogonal random variables. By using one of the almost sure approximation theorems (see Philipp (1986) for a pertinent survey), and Theorem 1 in Lacey and Philipp (1990) the same result holds for all kinds of dependent structures including some martingales, weak dependent random variables, etc.

Moreover, in remark (c), Lacey and Philipp (1990) noticed that the same method of proof, based on almost sure approximation, works for variables in the domain of normal attraction of a stable law.

By using a direct approach we give here a similar result for sequences in the domain of attraction of a stable law:

Theorem 1. Let $a_k = k^{-1}(\log k)^{\beta}$, $(\beta \geq 0,\ k \geq 1)$; $A_1 = 1$ and $A_j = \sum_{k=1}^{j} a_k$, for $j \geq 2$. Let $\{X_n\}$ be a sequence of i.i.d. random variables in the domain of attraction of a stable variable L with index α, $0 < \alpha \leq 2$, with the distribution function $G(x)$. (i.e. there are normalizing constants D_n and B_n such that $\frac{S_n - nD_n}{B_n} \xrightarrow[D]{} L$). Then, there is a P-null set N such that for $\omega \notin N$ and for all borel sets $A \subset R$ with $\lambda(\delta A) = 0$ we have

$$\lim_{n\to\infty} \frac{1}{A_n} \sum_{j=1}^{n} a_j I_A\left(\frac{S_j - jD_j}{B_j}\right) = P(L \in A) \tag{2}$$

□

For the construction of D_j, B_j, as well as for the characterization of distributions attracted by a stable law we refer to Feller, Vol. 11 (1971), Sections 9.6 and 17.5.

We notice that (2) has the following equivalent formulations.

For all $\omega \notin N$

$$\frac{1}{A_n} \sum_{j=1}^{n} a_j \delta\left(\frac{S_j - jD_j}{B_j}\right) \xrightarrow[D]{} L \quad \text{as } n \to \infty.$$

where $\delta(x)$ denotes the point mass at $x \in R$, or:

for all $\omega \notin N$, and for all functions f which are continuous and bounded:

$$\lim_{n\to\infty} \frac{1}{A_n} \sum_{j=1}^{n} a_j f\left(\frac{S_j - jD_j}{B_j}\right) = Ef(L).$$

The insight into the speed of convergence in Theorem A is given by some related central limit theorems in Weigl (1989) and Lacey and Philipp (1990).

A law of the iterated logarithms associated with Theorem 1 seems at this moment possible to get but we defer this subject for a forthcoming paper.

Here we shall apply Theorem 1 to analyze in detail some properties of the occupation time for a certain random walk. For this case the speed of convergence in Theorem 2 is studied and it is found the exact speed of this convergence.

Let $X_1, X_2, \ldots$ be a sequence of i.i.d.r.v.'s with

$$P\{X_1 = 1\} = P(X_1 = -1\} = \frac{1}{2}.$$

Define the random walk $\{S_n, n = 0,1,2,\ldots\}$ by

$$S_0 = 0, \quad S_n = X_1 + X_2 + \ldots + X_n \quad (n = 1,2,\ldots)$$

and consider the sequence $\{I(S_n), \ n = 0,1,2,\ldots\}$ where

$$I(t) = \begin{cases} 1 \text{ if } t \geq 0, \\ 0 \text{ if } t < 0. \end{cases}$$

Clearly

$$P\{I(S_n) = 1\} = \frac{1}{2} + O(n^{-1/2}).$$

Hence we might ask; does the strong law of large numbers hold i.e. we are interested in the limit behavior of $\sum_{k=0}^{n} I(S_k)$.

It is easy to see that

$$\limsup_{n\to\infty} \frac{1}{n} \sum_{k=0}^{n} I(S_k) = 1 \quad \text{a.s.} \tag{3}$$

and

$$\liminf_{n\to\infty} \frac{1}{n} \sum_{k=0}^{n} I(S_k) = 0 \quad \text{a.s.} \tag{4}$$

In fact (3) and (4) are simple consequences of

Theorem B. (Chung - Erdös, 1952). Let $f(x)$ be a nondecreasing function for which $\lim_{x\to\infty} f(x) = \infty$, $x/f(x)$ is nondecreasing and $\lim_{x\to\infty} x/f(x) = \infty$. Then

$$\sum_{k=1}^{n} I(S_k) \le n\left(1-\frac{1}{f(n)}\right) \text{ a.s.}$$

for all but finitely many n if and only if

$$\int_1^\infty \frac{dx}{x(f(x))^{1/2}} < \infty.$$

Consequence 1.

$$\sum_{k=1}^{n} I(S_k) \le n - \frac{n}{(\log n)^{2+\varepsilon}} \text{ a.s.}$$

for all but finitely many n if $\varepsilon > 0$. Further

$$\sum_{k=1}^{n} I(S_k) \ge n - \frac{n}{(\log n)^{2}} \text{ i.o.a.s.}$$

We have seen that the sequence $\{I(S_k)\}$ does not satisfy the strong law of large numbers. However from the Theorem A it follows

<u>Theorem C.</u> (Lévy, 1948)

$$\lim_{n\to\infty} \frac{1}{\log n} \sum_{k=1}^{n} \frac{I(S_k)}{k} = \frac{1}{2} \text{ a.s.}$$

i.e. the logarithmic density of $\{I(S_k)\}$ is 1/2.

It is natural to ask what happens if the sequence $\{1/k\}$ is replaced by another sequence $\{a_k\}$ in (3). In fact let $\{a_k\}$ be a monotone nonincreasing sequence of positive numbers with $\lim_{n\to\infty} a_n = 0$, $A_n = \sum_{k=1}^{n} a_k \to \infty (n\to\infty)$. Then we are interested in the limit behavior of $A_n^{-1} \sum_{k=1}^{n} a_k I(S_k)$. By remark (d) in Lacey and Philipp (1990) one can see that Theorem C fails for $a_k = k^{\varepsilon-1}$, $0 < \varepsilon \le 1$, due to arcsin law. As a consequence of Theorem A, we have in fact:

<u>Theorem 2.</u> Let $a_k = k^{\varepsilon-1} (0 < \varepsilon < 1,\ k = 1,2,\dots)$. Then

$$\limsup_{n\to\infty} A_n^{-1} \sum_{k=1}^{n} a_k I(S_k) = 1 \text{ a.s.}$$

and

$$\liminf_{n\to\infty} A_n^{-1} \sum_{k=1}^{n} a_k I(S_k) = 0 \quad \text{a.s.}$$

However as an application to Theorem 1 we have

<u>Theorem 3.</u> Let $a_k = k^{-1}(\log k)^{\alpha}(\alpha \geq 0,\ k = 3,4,\ldots)$. Then

$$\lim_{n\to\infty} A_n^{-1} \sum_{k=1}^{n} a_k I(S_k) = \frac{1}{2}.$$

Investigating the rate of convergence in Theorem C we present the following central limit theorem.

<u>Theorem D.</u> (Weigl, 1989).

$$\lim_{n\to\infty} P\left\{\frac{1}{c(\log n)^{1/2}} \left(\sum_{k=1}^{n} \frac{1}{k} I(S_k) - \frac{1}{2}\log n\right) < x\right\} = \Phi(x)$$

where

$$c = \left(\frac{1}{4\pi}\int_0^{\infty} A(y)dy\right)^{1/2}$$

and

$$A(y) = y^{-3/2}(\log(1+2y))^2 \qquad (0 < y < \infty).$$

A law of iterated logarithm type generalization of Theorems C, 3 and D is the following

<u>Theorem 4.</u> Let $a_k = k^{-1}(\log k)^{\alpha}$ $(\alpha \geq 0,\ k = 3,4,\ldots)$. Then

$$\limsup_{n\to\infty} \frac{1}{c_{\alpha}(\log n)^{\alpha+1/2}(\log_3 n)^{1/2}} \left(\sum_{k=1}^{n} a_k I(S_k) - \frac{1}{2}A_n\right) = 1 \quad \text{a.s.}$$

and

$$\liminf_{n\to\infty} \frac{1}{c_{\alpha}(\log n)^{\alpha+1/2}(\log_3 n)^{1/2}} \left(\sum_{k=1}^{n} a_k I(S_k) - \frac{1}{2}A_n\right) = -1 \quad \text{a.s.}$$

where

$$c_{\alpha} = \left(\frac{2^{-1}}{\pi(2\alpha+1)} \int_0^{\infty} A(y)dy\right)^{1/2}. \qquad \square$$

Here $\log_3 n$ denotes $\log(\log(\log n))$.

2. Proofs

In order to prove Theorem 1 we need some preliminary considerations.

(a) The result in Theorem 1 is equivalent to:

$$\lim_{n\to\infty} \frac{1}{A_n} \sum_{j=1}^{n} a_j f\left(\frac{S_j - jD_j}{B_j}\right) = Ef(L) \text{ a.s.} \tag{5}$$

for every function f(x) which is bounded, nondecreasing (or nonincreasing) and Lipschitz of any order ϑ, $0 < \vartheta \le 1$.

The proof of this fact is similar to the proof of Theorem 7.1 from Billingsley (1968) carried out with the function

$$\phi(t) = \begin{cases} 1 & \text{if } t \le 0 \\ 1-t & \text{if } 0 < t \le 1 \\ 0 & \text{if } t > 1 \end{cases}$$

Notice that this function has the desired properties and the family of functions constructed in the argument can be taken countable. In this way we can construct the P-null set N such that for all $\omega \notin N$ and for all $x \in R$

$$\lim_{n\to\infty} \frac{1}{A_n} \sum_{j=1}^{n} a_j I_{(-\infty,x]} \frac{S_j - jD_j}{B_j} = G(x)$$

and the result follows from the theory of weak convergence.

(b) The following lemma refers to the convergence of the moments in the central limit theorem (de Acosta and Giné, Theorem 6.1, 1979).

Lemma 1. Under the conditions of Theorem 1, for every $\beta \in (0,\alpha)$

$$\lim_{n\to\infty} E\left|\frac{S_n - nD_n}{B_n}\right|^{\beta} = \int |x|^{\beta} dG(x)$$

Proof of Theorem 1.

We denote by $Z_n = \frac{S_n - nD_n}{B_n}$. Because $Ef(Z_n) \to Ef(L)$ for every bounded continuous function f, by (a) we have to investigate if

$$Y_n = \frac{1}{A_n} \sum_{i=1}^{n} a_i (f(Z_i) - Ef(Z_i)) \to 0 \text{ a.s.}$$

for all functions f which are nondecreasing bounded and Lipschitz of any order γ, $0 < \gamma \le 1$.

We have the following estimate

$$A_n^2 EY_n^2 \le \sum_{i=1}^{n} a_i^2 \operatorname{var}(f(Z_i)) + 2 \sum_{j=1}^{n} \sum_{k=j+1}^{n} a_k a_j \operatorname{cov}(f(Z_k), f(Z_j))$$

The first sum is obviously convergent. In order to majorate the second sum we estimate the covariances. By independence and the properties of $f(x)$, for $k > j$ and $0 < \gamma < \min(\alpha, 1)$ we get:

$$\operatorname{cov}(f(Z_k), f(Z_j)) = E(f(Z_k) f(Z_j)) - Ef(Z_j) f\left(\frac{\sum_{i=j+1}^{k} X_i - jD_j + kD_k}{B_k}\right)$$

$$+ Ef(Z_j) Ef\left(\frac{\sum_{i=j+1}^{k} X_i - jD_j + kD_k}{B_k}\right) - Ef(Z_k) Ef(Z_j) = 0\left(E\left|\frac{\sum_{i=1}^{j} X_i - jD_j}{B_k}\right|^{\gamma}\right)$$

By Lemma 1 and the above considerations we get:

$$\operatorname{cov}(f(Z_j), f(Z_k)) = 0\left(\frac{B_j^{\gamma}}{B_k^{\gamma}}\right) \tag{6}$$

By the theory of the domains of attractions, B_n has the representation $n^{1/\alpha} h(n)$ where $h(n)$ is a function slowly varying at ∞. This fact implies:

$$\sum_{k=j+1}^{n} \frac{a_k}{B_k^{\gamma}} = 0\left(\frac{ja_j}{B_j^{\gamma}}\right)$$

This last relation and (6) prove that

$$A_n^2 EY_n^2 = 0\left(\sum_{i=1}^{n} a_i^2 + \sum_{j=1}^{n} ja_j^2\right) = 0\left(\sum_{j=1}^{n} ja_j^2\right)$$

So $EY_n^2 = 0\left(\frac{1}{\log n}\right)$.

Let $n_j = [\exp(j^{1+\varepsilon})]$ $(j = 1, 2, \ldots, \varepsilon > 0)$.

Then, by Borel-Cantelli lemma

$$\lim_{j\to\infty} Y_{n_j} = 0 \quad \text{a.s.}$$

Let $n_j \le n < n_{j+1}$. Then

$$|Y_n| \le |Y_{n_j}| + \frac{1}{A_{n_j}}\Big|\sum_{k=n_j+1}^{n_{j+1}} a_k\Big| \le |Y_{n_j}| + O\left(\frac{(j+1)^{(\beta+1)(1+\varepsilon)} - j^{(\beta+1)(1+\varepsilon)}}{j^{(\beta+1)(1+\varepsilon)}}\right)$$

Hence we obtain the statement.

The idea of proving Theorem 4 is to analyze a subsequence of $Y_n = \sum_{k=1}^{n} a_k I(S_k) - \frac{1}{2}A_n$, at moments of time given by successive visits of the random walk to 0. This subsequence is then decomposed into a martingale and an a.s. negligible sequence. Finally, a known LIL is applied to the resulting martingale.

The proof of Theorem 4 requires several lemmas.

<u>Lemma 2.</u>

$$g(z) = g_m(z) = \sum_{n=1}^{\infty} \frac{1}{n2^{2n-1}} \binom{2n-2}{n-1} (\log(1+\frac{2n}{z}))^m \quad (m > 1/2). \tag{7}$$

Then

$$\lim_{z\to\infty} z^{1/2} g(z) = \sqrt{\frac{2}{\pi}} \int_0^{\infty} A(y)dy$$

where

$$A(y) = A_m(y) = y^{-3/2}(\log(1+2y))^m. \tag{8}$$

<u>Proof.</u> Clearly

$$z^{1/2}g(z) = \sum_{n=0}^{\infty} z^{1/2} \frac{\sqrt{n+1}}{2^{2n+1}}\binom{2n}{n}\frac{1}{(n+1)^{3/2}}(\log(1+2\frac{n+1}{z}))^m = \sum_{n=0}^{\infty} b_{n+1} A\left(\frac{n+1}{z}\right)\frac{1}{z}$$

where

$$b_{n+1} = \frac{\sqrt{n+1}}{2^{2n+1}} \binom{2n}{n}.$$

Since by Stirling formula $\lim_{n\to\infty} b_n = \sqrt{\frac{2}{\pi}}$. Because $A(y)$ is a uniformly continuous function, integrable in $(0,\infty)$ and nonincreasing in (a,∞) if a is big enough we obtain the Lemma.

<u>Lemma 3.</u> Let

$$h(z) = h_m(z) = \sum_{n=1}^{\infty} c_n((\log(z+2n))^m - (\log z)^m)^2 \tag{9}$$

where

$$c_n = \frac{1}{n2^{2n-1}} \binom{2n-2}{n-1}.$$

Then for any $m > 0$

$$\lim_{z\to\infty} z^{1/2}(\log z)^{2-2m} h_m(z) = m^2\sqrt{\frac{2}{\pi}} \int_0^{\infty} A_2(y)dy.$$

<u>Proof.</u> Clearly

$$\begin{aligned} h(z) &= \sum_{n=1}^{\infty} c_n((\log z + \log(1+\frac{2n}{z}))^m - (\log z)^m)^2 \\ &= (\log z)^{2m} \sum_{n=1}^{\infty} c_n\left(\left(1 + \frac{\log(1+\frac{2n}{z})}{\log z}\right)^m - 1\right)^2 \\ &= (\log z)^{2m} \sum_{n=1}^{\infty} c_n\left(\sum_{k=1}^{\infty} \binom{m}{k} \left(\frac{\log(1+\frac{2n}{z})}{\log z}\right)^k\right)^2. \end{aligned}$$

By Lemma 2 we obtain

$$\sum_{n=1}^{\infty} c_n m^2 \frac{(\log(1+\frac{2n}{z}))^2}{(\log z)^2} = \frac{m^2}{(\log z)^2} g_2(z) \approx \frac{m^2}{(\log z)^2} z^{-1/2} \sqrt{\frac{2}{\pi}} \int_0^{\infty} A_2(y)dy$$

and for any $L = 3,4,...$

$$\sum_{n=1}^{\infty} c_n \frac{(\log(1+\frac{2n}{z}))^L}{(\log z)^L} = \frac{1}{(\log z)^L} g_L(z) \approx \frac{1}{(\log z)^L} z^{-1/2} \sqrt{\frac{2}{\pi}} \int_0^{\infty} A_{2L}(y)dy$$

Hence

$$h(z) \approx (\log z)^{2m} \frac{m^2}{(\log z)^2} z^{-1/2} \sqrt{\frac{2}{\pi}} \int_0^\infty A_2(y)dy$$

and Lemma 3 is proved.

The proof of the following lemma is nearly the same as that of Lemma 3. Hence it will be omitted.

Lemma 4. Let

$$\chi(z) = \chi_m(z) = \sum_{n=1}^{\infty} c_n((\log(z+2n))^m - (\log z)^m)^4. \qquad (10)$$

Then

$$\lim_{z\to\infty} z^{1/2}(\log z)^{4-4m}\chi_m(z) = m^4\sqrt{\frac{2}{\pi}} \int_0^\infty A_4(y)dy.$$

Let $\rho_0 = 0$ and

$$\rho_{n+1} = \min\{i : i > \rho_n, S_i = 0\}.$$

We list a few known properties of the r.v.'s ρ_n.

Lemma A.

$$P\{\rho_1 = 2\ell\} = \frac{1}{\ell 2^{2\ell-1}} \binom{2\ell-2}{\ell-1} \quad (\ell = 1,2,\dots) \qquad (11)$$

(see e.g. Rényi (1970) (4.8.29))

$$P\{\rho_k - k > n\} = P\{\max_{0\le j\le n} S_j < k\} = \sum_{\ell=0}^{k-1} \frac{1}{2^n} \binom{n}{[\frac{n-\ell}{2}]} \qquad (12)$$

(see e.g. Rényi (1970) (4.8.32) and (4.8.11))

$$\lim_{n\to\infty} P\{\frac{\rho_n}{n^2} < x\} = \frac{1}{\sqrt{2\pi}} \int_0^x \frac{e^{-1/2v}}{v^{3/2}} dv \quad (x > 0) \qquad (13)$$

(see e.g. Rényi (1970) (4.8.33))

$$\lim_{n\to\infty} \frac{1}{\log n} \sum_{k=1}^{n} \rho_k^{-1/2} = \pi^{-1/2} \text{ a.s.} \qquad (14)$$

(see Erdös-Taylor (1960)).

Applying the proving method of Erdös-Taylor the following slight generalization of (14) can be easily proved.

Lemma 5. For any $\alpha \ge 0$

$$\lim_{n\to\infty} \frac{1}{(\log n)^{\alpha+1}} \sum_{k=1}^{n} (\log \rho_k)^{\alpha} \rho_k^{-1/2} = \frac{2^{\alpha}}{(\alpha+1)\pi^{1/2}} \text{ a.s.}$$

Applying Lemma A by routine calculations one can obtain

Lemma 6.

$$\lim_{n\to\infty} P\{n\rho_n^{-1/2} < x\} = \sqrt{\frac{2}{\pi}} \int_0^x e^{-t^2/2} dt \qquad (0 < x < \infty) \tag{15}$$

and for any $\alpha \ge 0$

$$\lim_{n\to\infty} n^{2\alpha} E\rho_n^{-\alpha} = 2^{\alpha}\pi^{-1/2}\Gamma(\tfrac{1}{2}+\alpha), \tag{16}$$

and for any $\alpha \ge 0$, $\beta \ge 0$

$$\lim_{n\to\infty} n^{2\alpha}(\log n)^{-\beta} E(\log \rho_n)^{\beta} \rho_n^{-\alpha} = 2^{\alpha+\beta}\pi^{-1/2}\Gamma(\tfrac{1}{2}+\alpha). \tag{17}$$

Lemma 7. For any $m > 0$ we have

$$\lim_{n\to\infty} \frac{\rho_n^{1/2}}{(\log \rho_n)^{2m-2}} E(((\log \rho_{n+1})^m - (\log \rho_n)^m)^2 | \rho_n)$$

$$= m^2 \sqrt{\frac{2}{\pi}} \int_0^{\infty} A_2(y) dy.$$

Proof. By (11) we get

$$E(((\log \rho_{n+1})^m - (\log \rho_n)^m)^2 | \rho_n)$$

$$= \sum_{\ell=1}^{\infty} \frac{1}{\ell 2^{2\ell-1}} \binom{2\ell-2}{\ell-1} ((\log(\rho_n+2\ell))^m - (\log \rho_n)^m)^2$$

$$= h_m(z).$$

Hence Lemma 7 follows from Lemma 3.

Applying Lemma 4 instead of Lemma 3 we obtain

Lemma 8.

$$\lim_{n\to\infty} \frac{\rho_n^{1/2}}{(\log \rho_n)^{4m-4}} E(((\log \rho_{n+1})^m - (\log \rho_n)^m)^4 | \rho_n)$$

$$= m^4 \sqrt{\frac{2}{\pi}} \int_0^\infty A_4(y)dy.$$

Lemmas 6, 7 and 8 combined imply.

Lemma 9. For any $m > 0$

$$\lim_{n\to\infty} \frac{n}{(\log n)^{2m-2}} E((\log \rho_{n+1})^m - (\log \rho_n)^m)^2 = \frac{2^{2m-1}}{\pi} m^2 \int_0^\infty A_2(y)dy,$$

$$\lim_{n\to\infty} \frac{n}{(\log n)^{4m-4}} E((\log \rho_{n+1})^m - (\log \rho_n)^m)^4 = \frac{2^{4m-3}}{\pi} m^4 \int_0^\infty A_4(y)dy.$$

Proof of Theorem 4

The proof of Theorem 4 is based on the following

THEOREM OF HYDE AND SCOTT (1973, Corollary 1). Let $M_1, M_2, \ldots$ be a martingale difference sequence i.e.

$$E(M_n | M_{n-1}, \ldots, M_1) = 0 \quad \text{a.s.} \tag{18}$$

with

$$\mathcal{M}_n = \sum_{i=1}^n M_i,$$

$$E\mathcal{M}_n^2 = s_n^2 \to \infty \ (n\to\infty), \tag{19}$$

$$\sum_{n=1}^\infty s_n^{-4} E M_n^4 < \infty, \tag{20}$$

$$\lim_{n\to\infty} s_n^{-2} \sum_{k=1}^{n} M_k^2 = 1 \quad \text{a.s.} \tag{21}$$

Then

$$\limsup_{n\to\infty}(2s_n^2 \log\log s_n^2)^{-1/2} \mathcal{M}_n = 1 \quad \text{a.s.} \tag{22}$$

Let $\varepsilon_k = X_{\rho_{k-1}+1} (k = 1,2,\dots)$ i.e. $\varepsilon_k = 1$ if the k-th excursion of the random walk is positive, -1 otherwise. Further let

$$M_n = \varepsilon_n \sum_{i=\rho_{n-1}+1}^{\rho_n - 1} a_k = \varepsilon_n (A_{\rho_n - 1} - A_{\rho_{n-1}})$$

and

$$Y_n = \sum_{k=1}^{n} a_k I(S_k) - \frac{1}{2} A_n$$

$$\text{Then } Y_{\rho_n} = \frac{1}{2} \sum_{j=1}^{n} M_j + \sum_{i=1}^{n} a_{\rho_i}$$

Let $\mathcal{M}_n = \frac{1}{2} \sum_{j=1}^{n} M_j$ and observe that the sequence $\{M_n\}$ clearly satisfies (18). Further since

$$A_n \approx \frac{1}{\alpha+1}(\log n)^{\alpha+1}$$

we obtain

$$s_n^2 = E\mathcal{M}_n^2 \approx \frac{1}{4} \sum_{j=1}^{n} E\left(\frac{(\log \rho_j)^{\alpha+1} - (\log \rho_{j-1})^{\alpha+1}}{\alpha+1}\right)^2$$

where by Lemma 9

$$E\left(\frac{(\log \rho_j)^{\alpha+1} - (\log \rho_{j-1})^{\alpha+1}}{\alpha+1}\right)^2 \approx \frac{(\log j)^{2\alpha}}{j} \frac{2^{2\alpha+1}}{\pi} \int_0^\infty A_2(y)dy.$$

Hence we obtain (19) with

$$s_n^2 \approx \left(\frac{2^{2\alpha-1}}{\pi(2\alpha+1)} \int_0^\infty A_2(y)dy\right)(\log n)^{2\alpha+1} \tag{23}$$

Since by Lemma 9

$$EM_n^4 = \frac{1}{(\alpha+1)^4} E((\log (\rho_n - 1))^{\alpha+1} - (\log (\rho_{n-1})^{\alpha+1})^4$$

$$\approx \frac{(\log n)^{4\alpha}}{n} \frac{2^{4\alpha+1}}{\pi} \int_0^\infty A_4(y)dy$$

we obtain (20).

In order to get (21) we prove the following

Lemma 10.

$$\lim_{n\to\infty} \frac{\sum_{j=1}^n \left(M_j^2 - \frac{(\log \rho_j)^{2\alpha}}{\rho_j^{1/2}} \sqrt{\frac{2}{k}} \int_0^\infty A_2(y)dy\right)}{(\log n)^{2\alpha+1}} = 0 \quad \text{a.s.}$$

Proof. By Lemmas 7, 8 and 9 we easily get

$$E\left(\frac{\sum_{j=1}^n \left(M_j^2 - \frac{(\log \rho_j)^{2\alpha}}{\rho_j^{1/2}} \sqrt{\frac{2}{\pi}} \int_0^\infty A_2(y)dy\right)}{(\log n)^{2\alpha+1}}\right)^2 =$$

$$= 0\left(\frac{1}{(\log n)^{4\alpha+2}} \sum_{j=1}^n \frac{(\log j)^{4\alpha}}{j}\right) = 0\left(\frac{1}{\log n}\right).$$

Hence Lemma 10 follows by the usual arguments.

Having Lemmas 10, 5 and (23) we obtain (21). Consequently (22) holds for M_n. Now the result follows by the properties of ρ_n listed in Lemma 6.

References

de Acosta, A. - Giné, E. (1979) Convergence of Moments and Related Functionals in the General Central Limit Theorem in Banach Spaces. Z. Wahrsch. Verw. Gebiete 48, 213-231.

Billingsley, P. (1968) Convergence of Probability Measures. Wiley, New York.

Brosamler, G.A. (1988) An almost everywhere central limit theorem. Math. Proc. Camb. Phil. Soc. 104, 561-574.

Chung, K.L. - Erdös, P. (1952) On the application of the Borel-Cantelli lemma. Trans. Am. Math. Soc. 72, 179-186.

Erdös, P. - Taylor, S.J. (1960) Some problems concerning the structure of random walk paths. Acta Math. Acad. Sci. Hung. 11, 137-162.

Fisher, A., (1989) A Pathwise Central Limit Theorem for Random Walks (to appear in Annals of Probability).

Heyde, C.C. - Scott, D.J. (1973) Invariance principles for the law of iterated logarithm for martingales and processes with stationary increments. The Annals of Probability 1, 428-436.

Lacey, M. - Philipp, W. (1990) A note on the almost everywhere central limit theorem. Statistics and Probability Letters 3, 201-207.

Lévy, P. (1948) Processus Stochastiques et Mouvements Brownien. Gauthier Villars.

Newman, C.M. - Wright, A.L. (1982) Associated Random Variables and Martingale Inequalities. Z. Wahrsch. verw. Gebiete 59, 361-371.

Rényi, A. (1970) Foundations of Probability. Holden - Day San Francisco.

Schatte, P. (1988) On strong versions of the central limit theorems. Math. Nachr. 137, 249-256.

Weigl, A. (1989) Zwei Sätze über die Belegungszeit beim Random Walk. To appear.

UNIVERSALLY BAD SEQUENCES IN ERGODIC THEORY

by

Joseph Rosenblatt *

Mathematics Department

The Ohio State University

Columbus, Ohio 43210

Abstract

Different types, and different degrees, of universally bad sequences in ergodic theory are described. The main result is that for any sequence (μ_N) of probability measures on Z such that $\lim_{N\to\infty} \mu_N(k) = 0$ for all $k \in Z$, if there exists $b > 0$ and a dense subset $D \subset \{\gamma : |\gamma| = 1\}$ with $\liminf_{N\to\infty} |\hat{\mu}_N(\gamma)| \geq b$ for all $\gamma \in D$, then for any ergodic dynamical system $(X, \mathcal{B}, m, \tau)$, there is some $E \in \mathcal{B}$ such that the $\lim_{N\to\infty} \sum_{k=-\infty}^{\infty} \mu_N(k) 1_E(\tau^k x)$ fails to exist for all x in a set of positive measure.

* Partially supported by NSF Grant DMS-8802126

ISBN 0-12-085520-8

A dynamical system $(X, \mathcal{B}, m, \tau)$ will be a non-atomic separable probability space $(X, \mathcal{B}, m)$ equipped with an invertible measure-preserving transformation τ of $(X, \mathcal{B}, m)$. Krengel [15] proved that there exists a sequence of integers (n_k) which is universally bad in the sense that for all ergodic dynamical systems and $\epsilon > 0$, there exists $E \in \mathcal{B}$ such that $m(E) < \epsilon$ and $\limsup_{N\to\infty} \frac{1}{N} \sum_{k=1}^{N} 1_E(\tau^{n_k} x) = 1$ a.e.. Later Bellow [2] proved that any lacunary sequence (n_k) is universally bad in the sense that for all ergodic dynamical systems, there exists $f \in L_1(X)$ such that $\lim_{N\to\infty} \frac{1}{N} \sum_{k=1}^{N} f(\tau^{n_k} x)$ fails to exist for all x in a set of positive measure. Bellow's argument in [2] actually shows that for any $p, 1 \leq p < \infty$, the same holds for some $f \in L_p(X)$. Bellow, Jones, and Losert have also shown that lacunary sequences, $(a^k : k = 1, 2, 3, \ldots), a = 2, 3, \ldots,$ are strongly sweeping out; that is, these types of lacunary sequences are universally bad in the sense of Krengel's examples in [15]. The concept of strong sweeping out (see Definition 4) was first systematically used by Bellow. Also, see Bellow [1] for more on universally bad sequences.

By using a different argument than the ones in [2] and [15], specifically Bourgain's entropy theorem in [9], results that are in some ways stronger than those in [2] and [15] can be obtained. To clarify these results, consider these definitions.

1. Definition. A sequence (n_k) is L_p *universally bad* if for all ergodic dynamical systems, there is $f \in L_p(X)$ such that $\lim_{N\to\infty} \frac{1}{N} \sum_{k=1}^{N} f(\tau^{n_k} x)$ fails to exist for all x in a set of positive measure. The sequence is L_p *universally good* if for all ergodic dynamical systems and all $f \in L_p(X)$, $\lim_{N\to\infty} \frac{1}{N} \sum_{k=1}^{N} f(\tau^{n_k} x)$ exists a.e..

It is not hard to see, using the Conze principle [11] and the Banach principle of Sawyer that for p, $1 \leq p < \infty$, (n_k) is L_p universally bad if and only if there is no constant $C < \infty$ such that for all ergodic dynamical systems and all $f \in L_p(X)$, $m\{\sup_{N\geq 1} |\frac{1}{N} \sum_{k=1}^{N} f(\tau^{n_k} x)| > \lambda\} \leq \frac{C}{\lambda} \|f\|_p$. This is why the definition here when $p = 1$ is the same as the one in Bellow [2]. Also, because of the examples in Bellow [4], for any $p, 1 < p < \infty$, there exists (n_k) which is L_p universally bad, but L_q universally good for all $p < q \leq \infty$. It is not known if there is a sequence (n_k) which is

L_1 universally bad, but L_p universally good for all p, $1 < p \leq \infty$; nor is it known if there is a sequence which is L_p universally bad for all $p, 1 \leq p < \infty$, but which is L_∞ universally good.

Some universally bad sequences are not beyond redemption. For those that are, there is

2. Definition. A sequence (n_k) is L_p *persistently universally bad* if for all sequences (N_s), $N_s < N_{s+1}$ for $s \geq 1$, and all ergodic dynamical systems, there is $f \in L_p(X)$ such that $\lim_{s\to\infty} \frac{1}{N_s} \sum_{k=1}^{N_s} f(\tau^{n_k}x)$ fails to exist for all x in a set of positive measure.

For example, it is easy to construct a sequence (n_k) so that for two disjoint sequences (M_s) and (N_s), the discrete measures on Z, $\mu_s = \frac{1}{M_s}\sum_{k=1}^{M_s} \delta_{n_k}$ and $\nu_s = \frac{1}{N_s}\sum_{k=1}^{N_s} \delta_{n_k}$, satisfy

$$a) \sum_{s=1}^{\infty} \| \mu_s - \frac{1}{M_s}\sum_{k=1}^{M_s} \delta_k \|_1 < \infty, \text{and}$$

$$b) \sum_{s=1}^{\infty} \| \nu_s - \frac{1}{N_s}\sum_{k=1}^{N_s} \delta_{2k} \|_1 < \infty.$$

Then (n_k) is L_∞ universally bad (see Theorem 23), but for all dynamical systems and all $f \in L_p(X)$, $1 \leq p \leq \infty$, $\lim_{s\to\infty} \frac{1}{M_s}\sum_{k=1}^{M_s} f(\tau^{n_k}x)$ exists a.e.. So (n_k) is not L_1 persistently universally bad. This sequence (n_k) will have upper density 1 and lower density 0; also, the limit $\lim_{N\to\infty} \frac{1}{N}\sum_{k=1}^{N} f \circ \tau^{n_k}$ will not generally exist in any L_p norm. Another example, but now of density zero, is given by the sequence (η_k) composed of blocks $(2^m+1, \ldots, 2^m+m)$, $m \geq 1$. Bellow and Losert [3] show that this sequence is L_p universally bad, $1 \leq p < \infty$. But because of the lengthening blocks, if $\mu_N = \frac{1}{N}\sum_{k=1}^{N} \delta_{\eta_k}$, then $\hat{\mu}_N \to 0$ uniformly on each $S_\delta = \{\gamma : |\gamma| = 1, \delta \leq |\gamma - 1|\}$. Thus, the averages $\mu_N f = \frac{1}{N}\sum_{k=1}^{N} f \circ \tau^{\eta_k}$ converge in L_p norm, $1 \leq p < \infty$; but also, by Bellow, Jones, and Rosenblatt [6], there is a subsequence (N_s) such that $\lim_{s\to\infty} \mu_{N_s} f(x)$ exists a.e. for all $f \in L_p(X), 1 < p \leq \infty$. That is, (η_k) is not L_p persistently unversally bad for $1 < p \leq \infty$. It is not clear whether this sequence is L_1 persistently universally bad. Theorem 23 will show that there certainly are L_∞ persistently universally bad sequences.

Moreover, some L_∞ persistently universally bad sequences can be worse than others. Using the notation of [5], if μ is a probability measure on Z, and $f \in L_p(X)$, let $\mu f(x) = \sum_{k=-\infty}^{\infty} \mu(k) f(\tau^k x)$. Here, a sequence (μ_N) of probability measures on Z is called *dissipative* if $\lim_{N\to\infty} \mu_N(k) = 0$ for all $k \in Z$.

3. Definition. A sequence (n_k) is L_p *inherently universally bad* if for all dissipative sequences (μ_N) with supports in $\{n_k\}$, and all ergodic dynamical systems, there is $f \in L_p(X)$ such that $\lim_{N\to\infty} \mu_n f(x)$ fails to exist for a set of x of positive measure.

For example, in Theorem 2.3 it is shown that $(n_k) = (2^k)$ is L_∞ inherently universally bad. But not all L_∞ persistently universally bad sequences are L_1 inherently universally bad. To see this, note that by the subsequence theorem in [5], there is a sequence of blocks $B_s = (m_s + 1, \ldots, m_s + n_s)$ with $m_s, n_s \to \infty$ as $s \to \infty$, and $\lim_{s\to\infty} \frac{n_s}{\log m_s} = 0$ such that for all $f \in L_1(X)$, $\lim_{s\to\infty} \frac{1}{n_s} \sum_{k=m_s+1}^{m_s+n_s} f(\tau^k x)$ exists a.e.. Take the lacunary sequence $(\eta_k) = (2^k)$ and add to it a subsequence of the sequence (B_s) which is chosen rarely enough so that the new sequence (ξ_k) has

$$\sum_{N=1}^{\infty} \frac{\#\{\xi_k \in \bigcup_{s=1}^{\infty} B_s : k \leq N\}}{N} < \infty.$$

Then (ξ_k) is L_∞ persistently universally bad because (η_k) is, but (ξ_k) is not L_1 inherently universally bad because of the presence of the blocks. Indeed, there is a dissipative sequence (μ_N) supported in (ξ_k) such that, for any dynamical system,

$$\lim_{N\to\infty} \mu_N f(x) \text{ exists a.e. for all } f \in L_1(X).$$

These three definitions show that there are levels of bad behavior among universally bad sequences. Another sense in which this occurs for L_∞ universally bad sequences is in the degree of oscillation exhibited by the averages $\frac{1}{N} \sum_{k=1}^{N} f(\tau^{n_k} x)$. It is clear that (n_k) is L_∞ universally bad, even persistently or inherently so, if and only if the corresponding failure of convergence occurs for a characteristic function f. A gauge of the degree of oscillation in the averages is given by the sweeping out concept.

4. Definition. Given $\delta > 0$, a sequence of probability measures (μ_N) is δ *sweeping out* if for all ergodic dynamical systems and all $\epsilon > 0$, there is $E \in \mathcal{B}$ such that $\mu(E) < \epsilon$ and $\limsup_{N\to\infty} \mu_n 1_E \geq \delta$ a.e.. The sequence is *strongly sweeping out* if it is 1 sweeping out. A sequence (n_k), is δ *sweeping out* (strongly sweeping out) if the sequence (μ_N), where $\mu_N = \frac{1}{N}\sum_{k=1}^{N} \delta_{n_k}$, is δ sweeping out (strongly sweeping out).

It is also convenient to extend Definitions 1 and 2 to the context above. A sequence (μ_N) is L_p *(persistently) universally bad* if for all ergodic dynamical systems (and all subsequences (N_s)), there exists $f \in L_p(X)$ such that $\lim_{N\to\infty} \mu_N f(x)$ $(\lim_{s\to\infty}$ [illegible] $(x))$ fails to exist for all x in a set of positive measure.

Examples later will show that (n_k) exists which is L_∞ universally bad, but which is only δ sweeping out for some $\delta \leq \frac{1}{2}$. However, it is not clear whether an L_∞ universally bad sequence must be δ sweeping out for some $\delta > 0$ (cf. Remark 29). On the other hand, by del Junco and Rosenblatt [12], if (μ_N) is strongly sweeping out, then for a residual subset $\mathcal{R}$ of $\mathcal{B}$ in the symmetry pseudo-metric, if $E \in \mathcal{R}$, then $\limsup_{n\to\infty} \mu_N 1_E = 1$ a.e. and $\liminf_{N\to\infty} \mu_n 1_E = 0$ a.e.. This is the worst possible oscillation that can occur, but in any case δ sweeping out always implies the failure of a.e. convergence.

5. Proposition *If (μ_N) is δ sweeping out, then for all $\delta' < \delta$ and $\epsilon > 0$, and all ergodic dynamical systems, there exists $E \in \mathcal{B}$, $m(E) < \epsilon$, and a set $X' \subset X$ with $m(X \backslash X') \leq \epsilon$ such that*

$$\limsup_{N\to\infty} \mu_N 1_E - \liminf_{N\to\infty} \mu_N 1_E \geq \delta'$$

for all $x \in X'$.

Proof. Choose E with $m(E) \leq \epsilon$ and $\limsup \mu_n 1_E \geq \delta$ a.e.. Generally, $\int \liminf_{N\to\infty} \mu_N 1_E \, dm \leq \liminf_{N\to\infty} \int \mu_N 1_E \, dm \leq \epsilon$. Thus, $m\{\liminf_{N\to\infty} \mu_N 1_E \geq \delta - \delta'\} \leq \frac{\epsilon}{\delta-\delta'}$. So by reducing $\epsilon > 0$, there is $E \in \mathcal{B}$, $m(E) \leq \epsilon$, such that $\limsup_{N\to\infty} \mu_N 1_E - \liminf_{N\to\infty} \mu_N 1_E \geq \delta - (\delta - \delta') = \delta'$ on a subset $X' \subset X$ with $m(X \backslash X') \leq \epsilon$. □

6. Corollary. *If a sequence is δ sweeping out, then it is L_∞ universally bad.*

7. Question. If (μ_N) is δ sweeping out, does every dynamical system have a residual set of $E \in \mathcal{B}$ such that $\limsup_{N\to\infty} \mu_N 1_E - \liminf_{N\to\infty} \mu_n 1_E \geq \delta$ a.e.?

With these preliminary remarks, the theorems to follow will be easier to state. The central idea in the sequel is to apply the main theorem in Bourgain [9]. Stating this theorem for the special case used here, let (μ_N) be a sequence of probability measures on $\mathbb{Z}$. For a given dynamical system, to say that the L_2 *entropy of* (μ_N) *is finite* means that for all $\beta > 0$, there is $M_\beta < \infty$ such that if $f \in L_2(X)$, $\|f\|_2 \leq 1$, and $N_1, \ldots, N_M \geq 1$ are distinct and such that $\|\mu_{N_i} f - \mu_{N_j} f\|_2 \geq \beta$ for all $i \neq j$, then $M \leq M_\beta$.

8. Theorem (Bourgain). *If (μ_N) is a sequence of probability measures on $\mathbb{Z}$ and $(X, \mathcal{B}, m, \tau)$ is an ergodic dynamical system, then the L_2 entropy of (μ_N) is finite whenever $\lim_{N\to\infty} \mu_N f(x)$ exists* a.e. *for all $f \in L_p(X)$ for some fixed p, $1 \leq p \leq \infty$.*

9. Remark. It is important later that an even stronger fact is proved in [9]. Namely, with the hypotheses above, the L_2 entropy of (μ_N) is finite if $\int \sup_{N\geq 1} |\mu_N f| \, dm \to 0$ as $\int |f| \, dm \to 0$ among $f \in L_\infty(X)$, $\|f\|_\infty \leq 1$.

In order to handle a few details later that give sweeping out results, consider the following definitions for a sequence (μ_N) of probability measures on $\mathbb{Z}$. Given a dynamical system and a sequence (μ_N) of probability measures, let $\epsilon > 0$ and define $D_{\tau,\epsilon}$ to be the supremum of

$\int \sup_{N\geq 1} |\mu_N f(x)|dm(x)$ taken over $f \in L_\infty(X)$, $|f| \leq 1$, $\int |f|dm \leq \epsilon$. Let $\Delta_{\tau,\epsilon}$ be the supremum of $\int \sup_{N\geq 1} \mu_N 1_E(x)dm(x)$ taken over $E \in \mathcal{B}$, $m(E) \leq \epsilon$.

10. Proposition. *The maximal quantities $D_{\tau,\epsilon}$ and $\Delta_{\tau,\epsilon}$ satisfy $D_{\tau,\epsilon} \geq \Delta_{\tau,\epsilon}$ and $\Delta_{\tau,\epsilon/\delta} \geq D_{\tau,\epsilon} - \delta$.*

Proof. The first inequality is trivial. For the second, notice that if $\delta > 0$ and $|f| \leq 1$, then

$$\begin{aligned}\int \sup_{N\geq 1} |\mu_N f(x)|dm(x) &\leq \int \sup_{N\geq 1} \mu_N |f| dm \\ &\leq \int \sup_{N\geq 1} \mu_N 1_{\{|f|\geq\delta\}} dm + \int \sup_{N\geq 1} \mu_N(\delta 1_{\{|f|<\delta\}})dm \\ &\leq \int \sup_{N\geq 1} \mu_N 1_{\{|f|\geq\delta\}} dm + \delta,\end{aligned}$$

and

$$\int |f|dm \geq \delta m\{|f| \geq \delta\}.$$

Hence, if $\int |f|dm \leq \epsilon$ and $\int \sup_{N\geq 1} |\mu_N f|dm \geq D_{\tau,\epsilon} - \epsilon'$, for some $\epsilon' > 0$, then $m\{|f| \geq \delta\} \leq \epsilon/\delta$ and $\int \sup_{N\geq 1} |\mu_N 1_{\{|f|\geq\delta\}}|dm \geq D_{\tau,\epsilon} - \epsilon' - \delta$. Thus, $\Delta_{\tau,\epsilon/\delta} \geq D_{\tau,\epsilon} - \delta$. □

11. Corollary. *We have $\lim_{\epsilon\to 0} D_{\tau,\epsilon} = 0$ if and only if $\lim_{\epsilon\to 0} \Delta_{\tau,\epsilon} = 0$.*

Proof. If $D_{\tau,\epsilon} \geq \delta_0$ for all $\epsilon > 0$, then with $\delta = \delta_0/2$ in Proposition 10, $\Delta_{\tau,\epsilon/\delta_0} \geq \delta_0/2$ for all $\epsilon > 0$. □

12. Remark. Both $\Delta_{\tau,\epsilon}$ and $D_{\tau,\epsilon}$ are decreasing as ϵ decreases and Proposition 10 shows $\lim_{\epsilon\to 0} \Delta_{\tau,\epsilon} = \lim_{\epsilon\to 0} D_{\tau,\epsilon}$.

The next lemma is just the Conze principle, see Conze [11], in a special form.

13. Proposition. *Given two dynamical systems $(X_1, \mathcal{B}_1, m_1, \sigma)$ and $(X, \mathcal{B}, m, \tau)$ where τ is an aperiodic invertible measure-preserving transformation, $D_{\sigma,\epsilon} \leq D_{\tau,\epsilon}$ and $\Delta_{\sigma,\epsilon} \leq \Delta_{\tau,\epsilon}$ for all $\epsilon > 0$.*

Proof. The arguments are similar, so one of them will suffice. Clearly, given τ and σ_1, some invertible measure-preserving transformation from $X_1 \to X$,

$$\begin{aligned}\Delta_{\tau,\epsilon} &= \sup\left\{\int \sup_{N\geq 1} \mu_N 1_E dm : m(E) \leq \epsilon\right\}\\ &= \sup\left\{\int \sup_{N\geq 1} \mu_N 1_{\sigma_1 E} dm : m(E) \leq \epsilon\right\}\\ &= \sup\left\{\int \sup_{N\geq 1} \sum_{k=-\infty}^{\infty} \mu_N(k) 1_E(\sigma_1^{-1}\tau^k x) dm(x) : m(E) \leq \epsilon\right\}\\ &= \sup\left\{\int \sup_{N\geq 1} \sum_{k=-\infty}^{\infty} \mu_N(k) 1_E(\sigma_1^{-1}\tau^k \sigma_1 x) dm(x) : m(E) \leq \epsilon\right\}\\ &= \Delta_{\sigma_1^{-1}\tau\sigma_1,\epsilon}\,.\end{aligned}$$

Given $\epsilon' > 0$, choose $E \in \mathcal{B}$ with $m(E) \leq \epsilon$, and $M \geq 1$, such that $\Delta_{\sigma,\epsilon} - \epsilon' \leq \int \sup_{1\leq N\leq M} \sum_{k=-\infty}^{\infty} \mu_N(k) 1_E \circ \sigma^k dm$. By Halmos [13], there exists (σ_s), invertible measure-preserving transformations $\sigma_s : X_1 \to X$ such that $\sigma_s^{-1}\tau\sigma_s \to \sigma$ in the weak topology. So

$$\int \sup_{1\leq N\leq M} \sum_{k=-\infty}^{\infty} \mu_N(k) 1_E \circ (\sigma_s^{-1}\tau\sigma_s)^k dm \to \int \sup_{1\leq N\leq M} \frac{1}{N} \sum_{k=-\infty}^{\infty} \mu_N(k) 1_E \circ \sigma^k dm$$

as $s \to \infty$. Thus, for sufficiently large s,

$$\begin{aligned}\Delta_{\tau,\epsilon} = \Delta_{\sigma_s^{-1}\tau\sigma_s} &\geq \int \sup_{1\leq N\leq M} \sum_{k=-\infty}^{\infty} \mu_N(k) 1_E \circ (\sigma_s^{-1}\tau\sigma_s)^k dm\\ &\geq \int \sup_{1\leq N\leq M} \frac{1}{N} \sum_{k=-\infty}^{\infty} \mu_N(k) 1_E \circ \sigma^k dm \; - \; \epsilon'\\ &\geq \Delta_{\sigma,\epsilon} - 2\epsilon'\,.\end{aligned}$$

Thus, $\Delta_{\tau,\epsilon} \geq \Delta_{\sigma,\epsilon}$. □

14. Remark. This shows that $\Delta_{\tau_1,\epsilon} = \Delta_{\tau_2,\epsilon}$ and $\mathcal{D}_{\tau_1,\epsilon} = \mathcal{D}_{\tau_2,\epsilon}$ if τ_1 and τ_2 are ergodic.

15. Proposition. *If there is a dynamical system $(X_1, \mathcal{B}_1, m_1, \sigma)$ with $\lim_{\epsilon\to 0} \mathcal{D}_{\sigma,\epsilon} \neq 0$ then there exists $\delta > 0$ such that (μ_N) is δ sweeping out.*

Proof. From Proposition 13, there is a $\delta > 0$ such that for all ergodic dynamical systems $(X, \mathcal{B}, m, \tau)$, and all $\epsilon > 0$, $\mathcal{D}_{\tau,\epsilon} \geq 2\delta$. By the proof of Corollary 11, we have $\Delta_{\tau,\epsilon} \geq \delta$ for all $\epsilon > 0$. We proceed with the analysis on the dynamical system $(X, \mathcal{B}, m, \tau)$.

For any $\delta' > 0$, $\epsilon > 0$, there exists $E \in \mathcal{B}$, $m(E) < \epsilon$, $\int \sup_{N\geq 1} \mu_N 1_E dm \geq \delta - \delta'$. Since $\epsilon > 0$ is arbitrary, it is easy to see that for any $M \geq 1$, there exists $E_M \in \mathcal{B}$, $m(E_M) \leq \epsilon/2^M$, with $\int \sup_{N\geq M} \mu_N 1_{E_M} dm \geq \delta - 2\delta'$. But then $E_0 = \bigcup_{M=1}^{\infty} E_M$ has $m(E_0) \leq \epsilon$ and

$$\begin{aligned}\int \limsup_{N\to\infty} \mu_N 1_{E_0} dm &= \lim_{M\to\infty} \int \sup_{N\geq M} \mu_N 1_{E_0} dm \\ &\geq \delta - 2\delta'\end{aligned}$$

Fix δ_0, $0 < \delta_0 < \delta - 2\delta'$. Then if $L_{\delta_0} = \{\limsup_{N\to\infty} \mu_N 1_{E_0} \geq \delta_0\}$,

$$\begin{aligned}\delta - 2\delta' &\leq \int \limsup_{N\to\infty} \mu_N 1_{E_0} dm \\ &\leq \int_{L_{\delta_0}} 1 dm + \int_{X\setminus L_{\delta_0}} \delta_0 dm .\end{aligned}$$

Hence, $m(L_{\delta_0}) \geq (\delta - 2\delta' - \delta_0)/(1 - \delta_0)$. If one lets $\delta' = \delta/8$ and $\delta_0 = \delta/4$, then, this shows that for all $\epsilon > 0$, there exists $E \in \mathcal{B}$, $m(E) \leq \epsilon$, with $m\{\limsup_{N\to\infty} \mu_N 1_E \geq \delta/4\} \geq (\delta/2)/(1 - \delta/4) \geq \delta/2$.

Since $\delta > 0$ is fixed and $\epsilon > 0$ is arbitrary, as in the sweeping out theorem, Theorem 3 in Bellow, Jones, Rosenblatt [5], we can show that for all $\epsilon > 0$ there exists $E \in \mathcal{B}$, $m(E) \leq \epsilon$, with $\limsup_{N\to\infty} \mu_N 1_E \geq \delta/4$ a.e.. □

These propositions give this variation on Bourgain's theorem.

16. Theorem. *Given a sequence* (μ_N) *of probability measures on* Z, *if there is an ergodic dynamical system* $(X_1, \mathcal{B}_1, m_1, \sigma)$ *for which the* L_2 *entropy of* (μ_N) *is not finite, then* (μ_N) *is* δ *sweeping out for some* $\delta > 0$.

Proof. Remark 9 shows $\lim_{\epsilon\to 0} \mathcal{D}_{\sigma,\epsilon} \neq 0$. Proposition 15 finishes the proof. □

17. Remark. It is possible to take a different viewpoint to the above transfer from a fact about one ergodic dynamical system to a fact about all ergodic dynamical systems. For a given (μ_N) and dynamical system $(X_1, \mathcal{B}_1, m_1, \sigma)$, let $M_{\sigma,\beta}$ be the supremum over all M in the definition of the L_2 entropy. As in the proof of Proposition 13, for any aperiodic dynamical system $(X, \mathcal{B}, m, \tau)$, if $\epsilon > 0$, then $M_{\tau,\beta} \geq M_{\sigma,\beta+\epsilon}$ Thus, if $M_{\sigma,\beta}$ is not finite for some $\beta > 0$, then there is some $\beta > 0$ such that $M_{\tau,\beta}$ is infinite for all ergodic dynamical systems $(X, \mathcal{B}, m, \tau)$. So Theorem 16 could be proved by applying the Conze principle at a different point in the argument.

Before proving the main result, consider this analog of a well-known metric theorem in the theory of uniformly distributed sequences.

18. Proposition. *Given a sequence of probability measures on* Z, *if* $\lim_{N\to\infty} \max_k \mu_N(k) = 0$, *then for some subsequence* (N_s), $\lim_{s\to\infty} \hat{\mu}_{N_s}(\gamma) = 0$ *for a.e.* γ.

Proof. Here $\hat{\mu}_N(\gamma) = \sum_{h=-\infty}^{\infty} \mu_N(k)\gamma^{-k}$, so $\int_T |\hat{\mu}_N(\gamma)|^2\, d\gamma = \sum_{k=-\infty}^{\infty} \mu_N(k)^2 \leq \max_k \mu_N(k)$. Choose (N_s) such that $\sum_{s=1}^{\infty} \max_k \mu_{N_s}(k) < \infty$. Then $\int_T \sum_{s=1}^{\infty} |\hat{\mu}_{N_s}(\gamma)|^2\, d\gamma = \sum_{s=1}^{\infty} \int_T |\hat{\mu}_{N_s}(\gamma)|^2\, d\gamma \leq \sum_{s=1}^{\infty} \max_k \mu_{N_s}(k) < \infty$. So $\sum_{s=1}^{\infty} |\hat{\mu}_{N_s}(\gamma)|^2$ converges to a finite value for a.e. γ, and for a.e. γ, $\lim_{s\to\infty} \hat{\mu}_{N_s}(\gamma) = 0$. □

19. Remark. In the case $\mu_N = \frac{1}{N}\sum_{k=1}^{N} \delta_{n_k}$ for some sequence $(n_k) \subset Z$, this result can be made stronger in the usual way to show $\lim_{N\to\infty} \hat{\mu}_N(\gamma) = 0$ a.e..

20. Theorem. *Suppose* (μ_N) *is a dissipative sequence and there exists* $b > 0$ *such that* $\liminf_{N\to\infty} |\hat{\mu}_N(\gamma)| \geq b$ *for all* $\gamma \in D$ *where* D *is a dense subset of* T. *Then* (μ_N) *is* δ *sweeping out for some* $\delta > 0$.

Proof. There is no loss in assuming $\lim_{N\to\infty} \max_k \mu_N(k) = 0$. Otherwise, because (μ_N) is dissipative, there exists a $c > 0$ and increasing sequences (n_k) and (N_k) such that $\mu_{N_k} \geq c\delta_{n_k}$ for all $k \geq 1$. Then for any ergodic dynamical system, if $\epsilon > 0$, by the sweeping out theorem of Ellis and Friedman [13], there is $E \in \mathcal{B}$ such that $m(E) < \epsilon$ and $\limsup_{k\to\infty} 1_{\tau^{n_k}E} = 1$ a.e.. Hence, $\limsup_{k\to\infty} \mu_{N_k} 1_E \geq c$ a.e. and $\limsup_{N\to\infty} \mu_N 1_E \geq c$ a.e..

So we can assume $\lim_{N\to\infty} \max_k \mu_N(k) = 0$ and Proposition 18 applies; thus, by passing to a subsequence, it may be assumed also that $\lim_{N\to\infty} \hat{\mu}_N(\gamma) = 0$ a.e.. Theorem 16 says that only one ergodic dynamical system in which the L_2 entropy is infinite is needed to conclude the proof. It will be shown that if $\sigma(x) = \theta + x \mod 1$ where θ is irrational, then $[0,1]$, with Lebesgue measure λ, with the transformation σ , is such a dynamical system.

First, fix $M \geq 1$ and let $(\sigma_{ij} : i = 1, \ldots, 2^M, j = 1, \ldots, M)$ be an arbitrary matrix of 0's and 1's. By induction it is possible to choose $\gamma_1, \ldots, \gamma_{2^M} \in T$ and $N_1, \ldots, N_M$ distinct such that

$$a) \ |\hat{\mu}_{N_j}(\gamma_i)| < b/4 \quad \text{if} \quad \sigma_{ij} = 0$$

$$b) \ |\hat{\mu}_{N_j}(\gamma_i)| > b/2 \quad \text{if} \quad \sigma_{ij} = 1$$

To see this, let $V \subset T$ be a dense set such that $\lim_{N\to\infty} \hat{\mu}_N(\gamma) = 0$ for all $\gamma \in V$. Recall $\liminf_{N\to\infty} |\hat{\mu}_N(\gamma| \geq b$ for all $\gamma \in D$, where D is a dense set in T. Choose γ_i^1 in V or D as σ_{i1} is 0 or 1, respectively. There exists $N_0 \geq 1$ such that if $N \geq N_0$ then $|\hat{\mu}_N(\gamma_i^1)| < b/4$ if $\sigma_{i1} = 0$ and $|\hat{\mu}_N(\gamma_i^1)| > b/2$ if $\sigma_{i1} = 1$. Thus, we can choose N_1 so that a), b) are satisfied for all (i,j) with $j = 1$, when γ_i^1 is used in place of γ_i. Assume now that $(\gamma_i^J : i = 1, \ldots, 2^M, J = 1, \ldots, M_0)$ and $N_1, \ldots, N_{M_0}$ exist so that for each $J \leq M_0$, $a), b)$ hold for all (i,j) with $j \leq J$ when γ_i^J replaces γ_i. Choose $\gamma_i^{M_0+1}$ in V or D as σ_{i,M_0+1} is 0 or 1. By continuity of the $\hat{\mu}_{N_i}$, $i = 1, \ldots, M_0$, and the density of V and D, $\gamma_i^{M_0+1}$ can also be chosen so that a), b) still hold for all (i,j), $j \leq M_0$, with $\gamma_i^{M_0+1}$ replacing $\gamma_i^{M_0}$. Then choose N_{M_0+1} so that a), b) hold for all (i,j) with $j = M_0 + 1$. This completes the induction step of the choice of $(\gamma_i^J : i = 1, \ldots, 2^M, J = 1, \ldots, M)$ such that a), b) hold for all (i,j) with $j \leq J$ when γ_i^J replaces γ_i. Let $\gamma_i = \gamma_i^M$ for $i = 1, \ldots, M$. This gives a), b) for all $(i,j), i = 1, \ldots, 2^M, j = 1, \ldots, M$.

Choose any $\gamma_0 \in T$ of ∞ order. Then $\{\gamma_0^\ell : \ell \in Z\}$ is dense in T. Choose $(\sigma_{ij} : i = 1, \ldots, 2^M, j = 1, \ldots, M)$ to be 0's and 1's in such way that if $\mathcal{L}_j = \{i : \sigma_{ij} = 1\}$, then $\#\mathcal{L}_j = 2^{M-1}$

and the $\mathcal{L}_j$ are independent in that $\#(\mathcal{L}_{j_1} \cap \mathcal{L}_{j_2}) = 2^{M-2}$ for all $j_1 \neq j_2$, $1 \leq j_1, j_2 \leq M$. Let (γ_i) and (N_i) satisfy a), b) above for this choice of (σ_{ij}). Then choose $\ell_1, \ldots, \ell_{2^M}$ so that $\gamma_0^{\ell_i}$ is sufficiently close to γ_i in T that a), b) still hold with $\gamma_0^{\ell_i}$ replacing γ_i throughout. Let $f(\gamma) = \frac{1}{\sqrt{2^M}} \sum_{i=1}^{2^M} \overline{\gamma}^{\ell_i}$ for all $\gamma \in T$. Then $\|f\|_2 = 1$ and, with respect to the dynamical system of T with Lebesgue measure, and $\sigma(\gamma) = \gamma_0 \gamma$,

$$\begin{aligned}
\mu_N f(\gamma) &= \sum_{k=-\infty}^{\infty} \mu_N(k) f(\gamma_0^k \gamma) \\
&= \frac{1}{\sqrt{2^M}} \sum_{i=1}^{2^M} \sum_{k=-\infty}^{\infty} \mu_N(k) \overline{\gamma_0}^{\ell_i k} \overline{\gamma}^{\ell_i} \\
&= \frac{1}{\sqrt{2^M}} \sum_{i=1}^{2^M} \hat{\mu}_N(\gamma_0^{\ell_i}) \overline{\gamma}^{\ell_i}.
\end{aligned}$$

Hence, for any N_{j_1}, N_{j_2},

$$\| \mu_{N_{j_1}} f - \mu_{N_{j_2}} f \|_2^2 = \frac{1}{2^M} \sum_{i=1}^{2^M} |\hat{\mu}_{N_{j_1}}(\gamma_0^{\ell_i}) - \hat{\mu}_{N_{j_2}}(\gamma_0^{\ell_i})|^2$$

by the orthonormality of the functions $\gamma \mapsto \overline{\gamma}^p$. But then if $N_{j_1} \neq N_{j_2}$,

$$\begin{aligned}
\| \mu_{N_{j_1}} f - \mu_{N_{j_2}} f \|_2^2 &\geq \frac{1}{2^M} \sum_{\mathcal{L}_{j_1} \cap \mathcal{L}_{j_2}^c} |\hat{\mu}_{N_{j_1}}(\gamma_0^{\ell_i}) - \hat{\mu}_{N_{j_2}}(\gamma_0^{\ell_i})|^2 \\
&\quad + \frac{1}{2^M} \sum_{\mathcal{L}_{j_2} \cap \mathcal{L}_{j_1}^c} |\hat{\mu}_{N_{j_1}}(\gamma_0^{\ell_i}) - \hat{\mu}_{N_{j_2}}(\gamma_0^{\ell_i})|^2 \\
&\geq \frac{1}{2^M} \#(\mathcal{L}_{j_1} \cap \mathcal{L}_{j_2}^c) \left(\frac{b}{4}\right)^2 + \frac{1}{2^M} \#(\mathcal{L}_{j_2} \cap \mathcal{L}_{j_1}^c) \left(\frac{b}{4}\right)^2 \\
&= \frac{1}{4}\left(\frac{b}{4}\right)^2 + \frac{1}{4}\left(\frac{b}{4}\right)^2 = \frac{1}{2}\left(\frac{b}{4}\right)^2.
\end{aligned}$$

Hence, for all $j_1 \neq j_2$, $\| \mu_{N_{j_1}} f - \mu_{N_{j_2}} f \|_2 \geq \frac{b}{4\sqrt{2}}$. Since $j_i, j_2 = 1, \ldots, M$ is allowed here and $M \geq 1$ is arbitrary, this shows that the entropy of (μ_N) on $(T, \mathcal{B}, m, \sigma)$ is infinite, completing the proof. □

Remark. Clearly D need only be dense in some $E \subset T$ with $\lambda(E) > 0$.

This theorem wil allow us to show that any lacunary sequence is δ sweeping out and L_∞ inherently universally bad. First,

21. Lemma. *Given $\alpha > 0$, there exists $r = r(\alpha)$ such that if $\inf_k \frac{n_{k+1}}{n_k} \geq r$, then for any choice of $\sigma_k = \pm 1$, there is a dense set $D \subset [0,1]$ such that for all $\theta \in D$, there is $k = k(\theta) \geq 1$ such* that if $k \geq k(\theta)$, $|\exp(2\pi i n_k\theta) - \sigma_k| \leq \alpha$.

Proof. Given ϕ with $\exp(2\pi i n_k\theta) = \sigma_k$, if $|\theta - \phi| \leq \frac{\alpha}{2\pi n_k}$, then $|\exp(2\pi i n_k\theta) - \sigma_k| \leq \alpha$. So $\{\theta : |\exp(2\pi i n_k\theta) - \sigma_k| \leq \alpha\}$ consists of closed intervals of the same length, a length which is at least $\frac{\alpha}{\pi n_k}$, periodically repeated with left endpoints a distance $\frac{1}{n_k}$ apart. So if $\frac{2}{n_{k+1}} \leq \frac{\alpha}{\pi n_k}$, then at least one entire period if $\exp(2\pi i n_{k+1}\theta)$ occurs inside each of the closed intervals on which $|\exp(2\pi i n_k\theta) - \sigma_k| \leq \alpha$. Thus, if $\frac{n_{k+1}}{n_k} \geq \frac{2\pi}{\alpha}$, there exists a decreasing sequence of closed sets $(C_k), C_k \subset [0,1]$, such that

a) $C_1 = \{\theta : |\exp(2\pi i n_1\theta) - \sigma_1| \leq \alpha\}$,

b) C_k is a disjoint union of closed intervals $I_{k,s}$, $s = 1, \ldots, N_k$,

c) $|\exp(2\pi i n_k\theta) - \sigma_k| \leq \alpha$ for $\theta \in I_{k,s}$, $s = 1, \ldots, N_k$,

d) each $I_{k,s}$ contains at least one $I_{k+1,s}$.

Thus, $D_1 = \bigcap_{k=1}^{\infty} C_k$ is a closed set such that for each $\theta \in D_1$, $|\exp(2\pi i n_k\theta) - \sigma_k| \leq \alpha$ for all $k \geq 1$. Moreover, $D_1 \cap I_{1,s} \neq \phi$ for each $s = 1, \ldots, N_1$.

Now carry out the above construction with $(n_k : k \geq K)$ in place of $(n_k : k \geq 1)$. This gives a corresponding set D_K such that D_K meets each constituent interval of $\{\theta : |\exp(2\pi i n_K\theta) - \sigma_K| \leq \alpha\}$ and such that for all $\theta \in D_k$, if $k \geq K$, then $|\exp(2\pi i\theta n_k) - \sigma_k| \leq \alpha$. Let $D = \bigcup_{k=1}^{\infty} D_K$. Then D is dense and for all $\theta \in D$, $|\exp(2\pi i n_k\theta) - \sigma_K)| \leq \alpha$ for all sufficiently large k. □

22. Remark. The above construction can be carried out to guarantee D consists only of irrationals. If one chooses all $\sigma_k = 1$ and $\alpha = 1/2$, this shows that for suitably lacunary (n_k), there exists

$D \subset T$, D dense, such that for all dissipative sequences supported in (n_k), if $\gamma \in T$, there is $N_\gamma \geq 1$ such that for $N \geq N_\gamma$, $|\hat{\mu}_N(\gamma)| \geq 1/4$.

23. Theorem. *Suppose (n_k) is a sequence in Z with $\inf_{k\geq 1} n_{k+1}/n_k > 1$. Then any dissipative sequence (μ_N) supported in (n_k) is δ sweeping out for some $\delta > 0$; so (n_k) is L_∞ inherently universally bad and δ sweeping out for some $\delta > 0$.*

Proof. Suppose (μ_N) is a dissipative sequence supported in (n_k). Let $\gamma = \inf_k n_{k+1}/n_k$ and choose $R \geq 1$ such that $\gamma^R \geq r\left(\frac{1}{2}\right)$ where $r(\alpha)$ is as in Lemma 21. Each subsequence $(\eta_k^t) = (n_{kR+t} : k \geq 1)$ where $t = 0, \ldots, R-1$ satisfies $\inf_k \frac{\eta_{k+1}^t}{\eta_k^t} \geq \gamma^R \geq r\left(\frac{1}{2}\right)$. Each $\mu_N = \sum_{t=0}^{R-1} \nu_{Nt}$ where ν_{Nt} are positive measures supported in (η_k^t). There is some $t_0 = 0, \ldots, R-1, c > 0$, and an infinite sequence (N_s) such that $\| \nu_{N_s t_0} \|_1 \geq c$ for all $s \geq 1$. If $\nu_s = \nu_{N_s t_0} / \| \nu_{N_s t_0} \|_1$, then (ν_s) is a dissipative sequence and Remark 22 shows that $\liminf_{s\to\infty} |\hat{\nu}_s(\gamma)| \geq \frac{1}{4}$ for all γ in a dense subset D of T. Theorem 20 then shows $(\nu_s : s \geq 1)$ is δ sweeping out, and thus (μ_N) is $c\delta$ sweeping out because $\mu_{N_s} \geq c\nu_s$ for all $s \geq 1$, for some $\delta > 0$ □

The technique of Theorem 20 also allows one to construct non-lacunary sequences which are still pervasively universally bad. Specifically, consider the following.

24. Theorem. *There exists an L_∞ inherently universally bad sequence (n_k) such that $\lim_{k\to\infty} \frac{n_{k+1}}{n_k} = 1$.*

Proof. Consider $\{2^m 3^n : \ m, n \geq 1\}$ ordered as an increasing sequence (n_k). It is easy to see that $\lim_{k\to\infty} \frac{n_{k+1}}{n_k} = 1$. Let (μ_N) be a dissipative sequence supported in (n_k). There are three cases.

First, there may be $N_0 \geq 1$, $c > 0$ and a sequence (N_s) such that $\mu_{N_s}(\{2^m 3^{N_0} : \ m \geq 1\}) \geq c$ for all $s \geq 1$. If so let $\omega_s = \mu_{N_s}|_{\{2^m 3^{N_0}:\ m\geq 1\}}$ and $\nu_s = \omega_s / \| \omega_s \|_1$. Then $\mu_{N_s} \geq c\mu_s$. By Theorem 23, (ν_s) is δ sweeping out and so (μ_N) is $c\delta$ sweeping out.

Second, there may be $M_0 \geq 1, c > 0$ and a sequence (N_j) such that $\mu_{N_s}(\{2^{M_0} 3^n : n \geq 1\}) \geq c$ for all $s \geq 1$. In this case as above, (μ_N) is δ sweeping out for some $\delta > 0$.

Finally, if the first two cases fail then $\mu_N(\{2^m 3^{N_0} : m \geq 1\} \cup \{2^{M_0} 3^n : n \geq 1\}) \to 0$ as $N \to \infty$ for all $M_0, N_0 \geq 1$. But then the support of μ_N is asymptotically contained in $L_k = \{3^m 2^n = m, n \geq K\}$ for any fixed K i.e. $\mu_N(L_K) \to 1$ as $N \to \infty$. Hence, for any hexatic rational $\theta = \frac{p}{6^q}, p, q \in Z, q \geq 1, \lim_{N\to\infty} \hat{\mu}_N(\exp(2\pi i\theta)) = 1$. Since the hexatic rationals are dense in $[0,1]$, Theorem 20 shows (μ_N) is δ sweeping out for some $\delta > 0$. □

25. Remark. More is proved above than is stated; it is shown that any dissipative sequence (μ_N) supported in (n_k) is δ sweeping out.

Roger Jones pointed out another example of a non-lacunary L_∞ universally bad sequence which will be useful here. Let $[\![\cdot]\!]$ denote the greatest integer function.

26. Proposition. *Let* $n_k = k[\![\log_2 k]\!]$ *for* $k \geq 1$. *Then* (n_k) *is* δ *sweeping out for some* $\delta > 0$.

Proof. Given an ergodic dynamical system $(X, \mathcal{B}, m, \tau)$, the averages $\mu_N 1_E(x) = \frac{1}{N}\sum_{k=1}^{N} 1_E(\tau^{n_k}x)$ satisfy $\mu_{2N-1}1_E(x) \geq \frac{1}{2N-1}\sum_{k=N}^{2N-1} 1_E(\tau^{n_k}x) \geq \frac{1}{2}d_N 1_E(x)$ where $d_N = \frac{1}{N}\sum_{k=N}^{2N-1}\delta_{n_k}$. So it suffices to show (d_N) is δ sweeping out. But if $N = 2^{2^M}, d_N = \frac{1}{N}\sum_{k=N}^{2N-1}\delta_{k2^M}$. Hence for all dyadic rational θ, $\lim_{M\to\infty}\hat{d}_{2^{2^M}}(\exp(2\pi i\theta)) = 1$. By Theorem 20, (d_N) is δ sweeping out for some $\delta > 0$. □

27. Remark. Here $(n_k) = (k[\![\log_2 k]\!])$ is strictly increasing and $\lim_{k\to\infty}\frac{n_{k+1}}{n_k} = 1$; so the above proof that (n_k) is δ sweeping out, and therefore L_∞ universally bad, gives another example like that of Theorem 24. However, it is not clear if this (n_k) is L_∞ persistently bad.

Generally, δ sweeping out does not imply strongly sweeping out as the following two examples will show.

28. Example a). To see this for sequences of probability measures, let $\nu_N = \frac{1}{N}\sum_{k=1}^{N}\delta_k$ and let $\nu_N = \frac{1}{N}\sum_{k=1}^{N}\delta_{2^k}$ for $N \geq 1$. Let $\omega_N = \frac{1}{2}\nu_N + \frac{1}{2}\mu_N$. Then $\omega_N \geq \frac{1}{2}\nu_N$ and Theorem 23 show that (ω_N) is δ sweeping out for some $\delta > 0$. However, such a δ cannot be larger than $\frac{1}{2}$; so

(ω_N) is not strongly sweeping out. Indeed, for all ergodic dynamical systems, if $f \in L_1(X)$, $\lim_{N\to\infty} \mu_N f(x) = \int f\,dm$ a.e.. So for ergodic dynamical systems, $\limsup_{N\to\infty} \omega_N 1_E(x) = m(E) + \limsup_{N\to\infty} \frac{1}{2}\nu_N 1_E \le m(E) + \frac{1}{2}$. This same method, with $\omega_N = (1-\epsilon)\mu_N + \epsilon\nu_N$ for some $\epsilon > 0$, gives an example of (ω_N) which is δ sweeping out, but only for $\delta \le \epsilon$.

Example b). The essence of Example a) can be captured for sequences (n_k) using the sequence $(q_k) = (k [\![\log_2 k]\!] : k \ge 4)$ in Proposition 26. Let (p_k) be the prime numbers. Bourgain [8] (see also Wierdl [16]) showed that for all dynamical systems, if $f \in L_2(X)$, $\lim_{N\to\infty} \frac{1}{N}\sum_{k=1}^{N} f(\tau^{p_k}x)$ exists a.e.. Now each q_k is composite, so $\{p_k\} \cap \{q_k\} = \phi$. Also, it is easy to see that $\#\{q_k \le N\} \sim \frac{N}{\log_2 N}$ as $N \to \infty$. But the prime number theorem says that $\#\{p_k \le N\} \sim \frac{CN}{\log_2 N}$ as $N \to \infty$ where $C = \log_2 e$. Let $\mathcal{N} = \{p_k\} \cup \{q_k\}$ and emunerate $\mathcal{N} = (n_k)$ as an increasing sequence. This (n_k) is δ sweeping out for some $\delta > 0$, but it is not strongly sweeping out. To see this write $\frac{1}{N}\sum_{k=1}^{N} \delta_{n_k} = \frac{N_1}{N}\frac{1}{N_1}\sum_{k=1}^{N_1} \delta_{p_k} + \frac{N_2}{N}\frac{1}{N_2}\sum_{h=1}^{N_2} \delta_{q_k}$ where $N_1 = \#\{p_k \le n_N\}$ and $N_2 = \#\{q_k \le n_N\}$. Then $N_1 + N_2 = N$ and so $\lim_{N\to\infty} \frac{N}{N_2} = 1 + \lim_{N\to\infty} \frac{N_1}{N_2} = 1 + C$. But then for any ergodic dynamical system, $\lim_{N\to\infty} \frac{1}{N}\sum_{k=0}^{N} 1_E(\tau^{p_k}x) = f^*(x)$ a.e. with $\int f^*\,dm = m(E)$. Thus, $\limsup_{N\to\infty} \frac{1}{N}\sum_{k=1}^{N} 1_E(\tau^{n_k}x) \le \frac{C}{1+C} f^*(x) + \frac{1}{1+C}$. Because $\int f^* dm = m(E)$, this shows (n_k) is at most $\frac{1}{1+C}$ sweeping out. It also shows that $\limsup_{N\to\infty} \frac{1}{N}\sum_{k=1}^{N} 1_E(\tau^{n_k}x) \ge \frac{1}{1+C}\limsup_{N\to\infty} \frac{1}{N_1}\sum_{k=1}^{N_1} 1_E(\tau^{q_k}x) = \frac{1}{1+C}\limsup_{N\to\infty} \frac{1}{N}\sum_{k=1}^{N} 1_E(\tau^{q_k}x)$. So by Proposition 26, (n_k) is δ sweeping out for some $\delta > 0$.

Question. If (n_k) is L_∞ inherently universally bad, is it strongly sweeping out?

The examples of Theorem 24 and Proposition 26 show that rapidity of growth of (n_k) is not particularly necessary for universally bad behavior. In this same direction, consider the following two theorems.

30. Theorem. *Given (d_N) with $\lim_{N\to\infty} d_N = \infty$ and any (s_N), let $\mu_N = \frac{1}{N}\sum_{k=1}^{N} \delta_{s_N + kd_N}$ for $N \ge 1$. Then (μ_N) is δ sweeping out for some $\delta > 0$.*

Proof. By Theorem 20, it suffices to show there is a dense set $D \subset T, b > 0$, and a sequence (N_s) such that $\liminf_{s\to\infty} |\hat{\mu}_{N_s}(\gamma)| \geq b$ for all $\gamma \in D$. Since $|\hat{\mu}_N(\gamma)| = \frac{1}{N}|\sum_{k=1}^{N} \gamma^{kd_N}|$, without loss of generality all s_N can be taken to be 0.

Now the geometric sum $A_N = \frac{1}{N}\sum_{k=1}^{N} z^{kd}$, $z = \exp(2\pi i\theta)$, for $z^d \neq 1$, has $|A_N| = \frac{1}{N}\frac{|z^{Nd}-1|}{|z^d-1|}$. Let $D_N = \{\theta : \text{for some } p = 0,\dots,d-1, \frac{1}{3dN} \leq |\theta - \frac{p}{d}| \leq \frac{1}{2dN}\}$. If $\theta \in D_N$, then $|d\theta - p| \leq \frac{1}{2N}$ for some p and so $|\exp(2\pi i d\theta) - 1| \leq \frac{\pi}{N}$. That is, $\frac{1}{N|z^d-1|} \geq \frac{1}{\pi}$. But $dN\theta$ cannot be too close to Z. Indeed if there is $P \in Z$ such that $|dN\theta - P| \leq \frac{1}{8}$, then $|d\theta - \frac{P}{N}| \leq \frac{1}{8N}$ and so, using $|d\theta - p| \leq \frac{1}{2N}$, it follows that $|p - \frac{P}{N}| \leq \frac{5}{8N}$. Hence, $|Np - P| \leq \frac{5}{8}$ and so $Np = P$. But then we can use $\frac{1}{3dN} \leq |\theta - \frac{p}{d}|$ for the same p as above, giving $\frac{1}{3} \leq |dN\theta - Np| = |dN\theta - P| \leq \frac{1}{8}$, a contradiction. Hence, $|dN\theta - P| > \frac{1}{8}$ for all $P \in Z$ and thus $|\exp(2\pi i dN\theta) - 1| = |z^{dN} - 1| \geq C$ for some universal constant ($C = 2\sin\left(\frac{1}{16}\right)$ will do). Thus, for $\theta \in D_N$, $|A_N(\theta)| \geq \frac{C}{\pi} \geq \frac{3}{100}$.

Because $\lim_{N\to\infty} d_N = \infty$, there is a subsequence (d_{N_s}) which is sufficiently lacunary that $\mathcal{D} = \bigcup_{s=1}^{\infty} \bigcap_{s=S}^{\infty} D_s$ is dense in $[0,1]$. Let $D = \{\exp(2\pi i\theta) : \theta \in \mathcal{D}\}$. Then $\liminf_{s\to\infty} |\hat{\mu}_{N_s}(\gamma)| \geq \frac{3}{100}$ for all $\gamma \in D$. □

31. Remark. a) The proof above can be carried out in any subsequence (N_t). Hence, actually any (μ_{N_t}) is δ sweeping out and so (μ_N) is L_∞ persistently universally bad.

b) One corollary of this result is that there cannot be any pointwise ergodic theorem for averages of the form $\frac{1}{N}\sum_{k=1}^{N} f(\tau^{kN}x)$.

32. Theorem. *Given any* (g_k) *with* $g_k \geq 1$ *and* $\lim_{k\to\infty} g_k = \infty$, *there exists* (n_k) *which is* δ *sweeping out, and thus* L_∞ *universally bad, such that* $n_{k+1} - n_k \leq g_k$ *for all* $k \geq 1$.

Proof. Choose any (d_N), $d_N \geq 1$, and then choose (s_N) so that the blocks of integers $B_N = (s_N + d_N, \dots, s_N + M_N d_N)$ are disjoint, but $s_N + M_N d_N + 1 = s_{N+1} + d_{N+1}$. By choosing $M_{N+1}/M_N \geq 2$ for all $N \geq 1$, $\# \bigcup_{N=1}^{N_0} B_N \leq \# B_{N_0+1}$. Let (n_k) be an enumeration of $\bigcup_{N=1}^{\infty} B_N$ in increasing order. Then $\frac{1}{M}\sum_{k=1}^{M} \delta_{n_k} \geq \frac{1}{2}\frac{1}{M_L}\sum_{k\in B_L} \delta_k$ when $\{n_1,\dots,n_M\} = \bigcup_{N=1}^{L} B_N$. If $\mu_L =$

$\frac{1}{M_L} \sum_{k \in B_L} \delta_k$, $L \geq 1$, then (μ_L) is δ sweeping out by Theorem 30 if $\lim_{N \to \infty} d_N = \infty$. Also, if (d_N) increasing slowly enough, then $n_{k+1} - n_k \leq g_k$ for all $k \geq 1$. Thus, it is possible to choose (d_N) so that both conditions are met. □

33. Remark. It is probably possible to construct (n_k) with $n_{k+1} - n_k$ growing as slowly as one likes, but such that (n_k) is L_∞ inherently universally bad. However, if $n_{k+1} - n_k$ remains bounded, then (n_k) is never L_1 universally bad. Indeed, this bound shows $\liminf_{N \to \infty} \frac{\#\{n_k \leq N\}}{N} > 0$. Hence, by the usual ergodic theorem, there is a weak maximal estimate $m\{x : \sup_{N \geq 1} \frac{1}{N} |\sum_{k=1}^{N} f(\tau^{n_k} x)| > \lambda\} \leq \frac{C}{\lambda} \| f \|_1$, for some constant C, for all $\lambda > 0$. But then for a.e. irrational θ, $(n_k\theta)$ is uniformly distributed mod 1. Letting $\tau(x) = (x + \theta) mod 1$, for all continuous $f : [0, 1] \to R$, $f(0) = f(1)$, $\lim_{N \to \infty} \frac{1}{N} \sum_{k=1}^{N} f(\tau^{n_k} x) = \int_0^1 f(s)\, ds$ uniformly in x. Thus, for all $f \in L_1[0, 1]$, $\lim_{N \to \infty} \frac{1}{N} \sum_{k=1}^{N} f(\tau^{n_k} x)$ exists a.e.. Hence, for sequences with bounded gaps, the right question is whether the sequence is L_1 universally good. The example in Bourgain [10] of a sequence (n_k) which is universally good for mean convergence, and has a positive density, but is not L_1 universally good, can be used to some purpose here. Indeed, let $\{\eta_k\} = 2Z \cup \{2n_k + 1 : k \geq 1\}$. Then as in Example 18b), (η_k) is not universally good, but it does have gaps $\eta_{k+1} - \eta_k \leq 2$. It would be very interesting to obtain a characterization of sequences (n_k) with bounded gaps which are L_1 universally good.

References

1. Bellow, A. (1982), Sur la structure des suites “mauvaises universelles” en theorie ergodique, Comptes Rendus 294, 55-58.

2. Bellow, A. (1983), On "Bad Universal" sequences in ergodic theory (II), Springer-Verlag Lecture Notes, #1033, 74–78.

3. Bellow, A. and Losert, V. (1985), The weighted pointwise ergodic theorem and the individual ergodic theorem along subsequences, Trans. AMS 288, 307-345.

4. Bellow, A. (1989), Perturbation of a sequence, Advances in Math, 78 no. 2, 131-139.

5. Bellow, A., Jones, R., and Rosenblatt, J. (1990), Convergence for moving averages, to appear in Dynamical Systems and Ergodic Theory.

6. Bellow, A., Jones, R. and Rosenblatt, J. (1990), Almost everywhere convergence of weighted averages, preprint.

7. Bellow, A., Jones, R. and Losert, V.(1990), The strong sweeping out property for lacunary sequences (r^k), for the Riemann sums and related matter, preprint.

8. Bourgain, J. (1987), An approach to pointwise ergodic theorems, Springer–Verlag Lecture Notes, Vol. 1317, 204-223, Springer-Verlag, Berlin-Heidelberg-New York, 1986-1987.

9. Bourgain, J. (1988), Almost sure convergence and bounded entropy, Israel J. Math 63, 79-97.

10. Bourgain, J. (1988), Return times of dynamical systems, preprint IHES.

11. Conze, J. P. (1973), Convergence des moyennes ergodiques pour des sous-suites, Bull. Soc. Math., France 35, 7–15.

12. del Junco, A. and Rosenblatt, J. (1979), Counterexamples in ergodic theory and number theory, Math. Ann 245, 185-197.

13. Ellis, M. and Friedman, N. (1978), Sweeping out on a set of integers, Proc. AMS 72 #3, 509-512.

14. Halmos, P. (1956) *Lectures in Ergodic Theory*, Chelsea Publishing Co., New York.

15. Krengel, V. (1971), On the individual ergodic theorem for subsequences, Ann. Math. Stat. 42 #3, 1091–1095.

16. Wierdl, M. (1988), Pointwise ergodic theorem along prime numbers, Israel J. of Math 64, 315-336.

On an Inequality of Kahane

Yoram Sagher and Kecheng Zhou

Let $(\mathbf{X}, \|\cdot\|)$ be a Banach space. Let $x(t)$ be a function from $[0,1]$ to $\mathbf{X}$. Let $I_{l,j} = [(j-1)2^{-l}, j2^{-l})$, $l = 0,1,2,\cdots . j = 1,2,\cdots,l$, be the dyadic intervals in $[0,1]$. Let $r_k(t), k = 0,1,\cdots$ be Rademacher functions defined as

$$r_0(t) = \begin{cases} 1 & 0 \le t < 1/2 \\ -1 & 1/2 \le t < 1 \end{cases},$$

$$r_0(t+1) = r_0(t),$$

and

$$r_k(t) = r_0(2^k t), \qquad k = 1,2,\cdots.$$

Define

$$x_I = \frac{1}{|I|}\int_I x(t)dt,$$

$$x^{\#}(t_0) = sup\{\frac{1}{|I|}\int_I \|x(t) - x_I\| dt\}$$

where supremum is taken over all the dyadic intervals I such that $t_0 \in I$, and

$$\|x\|_{BMO_d(\mathbf{X})} = esssup_{0 \le t \le 1} x^{\#}(t).$$

Denote the vector-valued dyadic BMO space as

$$BMO_d(\mathbf{X}) = \{x(t) : [0,1] \to \mathbf{X}, \ \|x\|_{BMO_d(\mathbf{X})} < \infty.\}$$

ISBN 0-12-085520-8

Theorem 1 *Let* $(\mathbf{X}, \|\cdot\|)$ *be a normed space and* $x_0, x_1, \cdots, x_n$ *be any vectors in* $\mathbf{X}$, *then* $\sum_{k=0}^{n} r_k(t)x_k \in BMO_d(\mathbf{X})$, *and*

$$\|\sum_{k=0}^{n} r_k(t)x_k\|_{BMO_d(\mathbf{X})} \leq \int_0^1 \|\sum_{k=0}^{n} r_k(t)x_k\|dt.$$

Proof. Let $x(t) = \sum_{k=0}^{n} r_k(t)x_k$. Let I be any dyadic interval of length 2^{-l}. Assume that $n \geq l$. Note that $r_k(t)$ is identically constant on I , denoted as $r_k(I)$, for $k < l$, and that $\int_I r_k(t)dt = 0$, for $k \geq l$. We have

$$x_I = \frac{1}{|I|}\int_I x(t)dt = \sum_{k<l} r_k(I)x_k.$$

Also note that if $n < l$, then $x_I = \sum_{k=0}^{n} r_k(I)x_k$.

Let $t_0 \in [0,1]$. Let I be any dyadic interval of length 2^{-l} such that $t_0 \in I$. Then, for $n \geq l$,

$$\frac{1}{|I|}\int_I \|x(t) - x_I\|dt = \frac{1}{|I|}\int_I \|\sum_{k\geq l} r_k(t)x_k\|dt.$$

By a change of variable, we have

$$\frac{1}{|I|}\int_I \|\sum_{k\geq l} r_k(t)x_k\|dt = \int_0^1 \|\sum_{k\geq l} r_{k-l}(t)x_k\|dt.$$

Notice that since the Rademacher functions are independent and identically distributed, $\|\sum_{k\geq l} r_{k-l}(t)x_k\|$ and $\|\sum_{k\geq l} r_k(t)x_k\|$ are identically distributed functions of t. We therefore have

$$\int_0^1 \|\sum_{k\geq l} r_{k-l}(t)x_k\|dt = \int_0^1 \|\sum_{k\geq l} r_k(t)x_k\|dt.$$

As a simple application of Billard's theorem (see [2]), we have

$$\int_0^1 \|\sum_{k=l}^{n} r_k(t)x_k\|dt \leq \int_0^1 \|\sum_{k=0}^{n} r_k(t)x_k\|dt.$$

Since the proof of Billard's theorem is easy in the special case we are interested in, we give the argument: Let $\epsilon_k = -1$, for $k < l$, and $\epsilon_k = 1$, for $k \geq l$. Then we have

$$\begin{aligned}\|\sum_{k=l}^{n} r_k(t)x_k\| &= \frac{1}{2}\|\sum_{k=0}^{n} r_k(t)x_k + \sum_{k=0}^{n} \epsilon_k r_k(t)x_k\| \\ &\leq \frac{1}{2}(\|\sum_{k=0}^{n} r_k(t)x_k\| + \|\sum_{k=0}^{n} \epsilon_k r_k(t)x_k\|)\end{aligned}$$

Note that $\|\sum_{k=0}^{n} r_k(t)x_k\|$ and $\|\sum_{k=0}^{n} \epsilon_k r_k(t)x_k\|$ are identically distributed. We have

$$\int_0^1 \|\sum_{k=l}^{n} r_k(t)x_k\|dt \leq \int_0^1 \|\sum_{k=0}^{n} r_k(t)x_k\|dt.$$

Hence we obtain that, if $n \geq l$, then

$$\frac{1}{|I|}\int_I \|x(t) - x_I\|dt \leq \int_0^1 \|\sum_{k=0}^{n} r_k(t)x_k\|dt.$$

If $n < l$, then the inequality above holds trivially because the left-hand side of the inequality is equal to zero. Therefore, we proved that

$$x^{\#}(t_0) \leq \int_0^1 \|\sum_{k=0}^{n} r_k(t)x_k\|dt$$

so that $\sum_{k=0}^{n} r_k(t)x_k \in BMO_d(\mathbf{X})$ and

$$\|\sum_{k=0}^{n} r_k(t)x_k\|_{BMOd(\mathbf{X})} \leq \int_0^1 \|\sum_{k=0}^{n} r_k(t)x_k\|dt.$$

The well-known theorem of John- Nirenberg (see [1]) concerning the distribution function of functions in BMO extends without change to vector valued BMO_d.

Theorem 2 *There exist constants B and b, so that for any $x(t) \in BMO_d(\mathbf{X})$, and for any dyadic interval $I \subset [0,1]$ and $\alpha > 0$,*

$$|\{t \in I : \|x(t) - x_I\| > \alpha\}| \leq B|I|exp\{-\frac{b\alpha}{\|x\|_{BMO_d(\mathbf{X})}}\}.$$

As an immediate consequence, we have

Theorem 3 *Given $0 < p < \infty$, and a dyadic interval I, let $x_0, x_1, \cdots, x_n$ be any vectors in $\mathbf{X}$, and denote $x(t) = \sum_{k=0}^{n} r_k(t)x_k$, Then we have:*

$$\left(\frac{1}{|I|}\int_I \|x(t) - x_I\|^p dt\right)^{1/p} \le B_p \int_0^1 \|\sum_{k=1}^{n} r_k(t)x_k\| dt,$$

where B_p is a constant which depends only on p.

Note that if $I = [0,1]$, then $x_I = 0$. This proves the inequality on the right in the following theorem (See [3]):

Theorem 4 *(Kahane's inequalities) For any $0 < p < \infty$ and any $x_0, x_1, \cdots, x_n$ in $\mathbf{X}$ we have:*

$$A_p \int_0^1 \|\sum_{k=0}^{n} r_k(t)x_k\| dt \le \left(\int_0^1 \|\sum_{k=0}^{n} r_k(t)x_k\|^p dt\right)^{1/p} \le B_p \int_0^1 \|\sum_{k=0}^{n} r_k(t)x_k\| dt.$$

Proof. We need to show only the left-hand side inequality for $0 < p < 1$.

Let $0 < p < 1$ and θ be such that $(1-\theta)/p + \theta/2 = 1$. Then

$$\begin{aligned}
\int_0^1 \|\sum_{k=0}^{n} r_k(t)x_k\| dt &= \int_0^1 \|\sum_{k=0}^{n} r_k(t)x_k\|^{1-\theta} \cdot \|\sum_{k=0}^{n} r_k(t)x_k\|^{\theta} dt \\
&\le \left(\int_0^1 \|\sum_{k=0}^{n} r_k(t)x_k\|^p dt\right)^{(1-\theta)/p} \left(\int_0^1 \|\sum_{k=0}^{n} r_k(t)x_k\|^2 dt\right)^{\theta/2} \\
&\le \left(\int_0^1 \|\sum_{k=0}^{n} r_k(t)x_k\|^p dt\right)^{(1-\theta)/p} B_2 \cdot \left(\int_0^1 \|\sum_{k=0}^{n} r_k(t)x_k\| dt\right)^{\theta}.
\end{aligned}$$

This implies the left-hand side inequality for $0 < p < 1$.

References

1. F. John and L. Nirenberg, On Functions of Bounded Mean Oscillation, Communications on Pure and Applied Mathematics, 14, 1961, pp 415-426.

2. J. P. Kahane, Some Random Series of Functions, Second Edition, Cambridge University Press, 1985.

3. V. D. Milman and G. Schechtman, Asymptotic Theory of Finite Dimensional Normed Spaces, Lecture Notes in Mathematics, No. 1200, 1986.

A PRINCIPLE FOR ALMOST EVERYWHERE CONVERGENCE OF MULTIPARAMETER PROCESSES *

Louis Sucheston and László I. Szabó

The Ohio State University, Department of Mathematics

Columbus, Ohio, USA

Introduction

The proofs of ergodic and martingale theorems are similar. The existence of a unified approach to martingale and ergodic theory is a well known old problem, explicitly stated in Doob's book [**9**]. A. and C. Ionescu Tulcea [**17**] gave a remarkable common generalization of vector-valued martingale and ergodic theorems. It has been known that the passage from weak to strong maximal inequalities can be done by a general argument applicable to harmonic analysis, ergodic theory, and martingale theory; for a general approach involving Orlicz spaces and their hearts, see [**13**]. In [**29**], a simple unified (martingale + ergodic theorems) passage from one to many parameters is given, based on a general argument valid for order-continuous Banach lattices; see also [**16**]. This approach gives a unified short proof of many known theorems, namely multiparameter versions of theorems of Doob (Cairoli's theorem [**4**] in stronger form, not assuming commutation), theorems of Dunford-Schwartz [**11**] and Fava [**14**]; the multiparameter point-transformation case was proved by Zygmund [**30**] and Dunford [**10**], also the theorems of Akcoglu [**1**], Stein [**28**], Rota [**27**].

---

* The research of the authors was supported in part by the National Science Foundation under grant DMS 88-02126.

ISBN 0-12-085520-8

In the present paper, we first develop in detail the principle that allows the reduction to one parameter. There is also a one-sided version of this principle, needed to obtain "demiconvergence" in many parameters; see also **[12]**, **[24]** and **[25]**. As an application, in Section 2 we consider martingales. It has been known that less than the independence assumption of Cairoli **[4]** is needed; if a martingale is also a "block-martingale" then the independence may be dispensed with, as was shown in **[12]**, **[15]** and **[24]**. Conversely, under the independence assumption every martingale is a block martingale. Here we reduce the case of block martingale to successive applications of the conditional expectation operator, which allows the application of the principle and greatly simplifies the proofs. In Section 3, we give a superadditive variant of the multiparameter Chacon-Ornstein theorem.

1. A Multiparameter Convergence Principle

Let E be a σ-complete Banach lattice, i.e., a Banach lattice such that every order bounded sequence has a least upper bound in E (see **[22]**). E is said to have an *order-continuous norm* if for every net (equivalently, sequence) (f_i), $f_i \downarrow 0$ implies $\|f_i\| \downarrow 0$. Let $F \subset E$. A map $T : F \to E$ is *increasing*, if $f \leq g$ implies $Tf \leq Tg$; *positive* if $f \geq 0$ implies $Tf \geq 0$; *linear* if $T(\alpha f + \beta g) = \alpha Tf + \beta Tg$ for any $\alpha, \beta \in \mathbf{R}$; *positively homogeneous* if $|T(\alpha f)| = |\alpha| Tf$ for each $f \in F$, $\alpha \in \mathbf{R}$; *subadditive* if $T(f+g) \leq Tf + Tg$. A map that is both positively homogeneous and subadditive is called *sublinear*. Sublinear increasing maps are positive since $0 \leq f$ implies $T0 = 0 \leq Tf$. A map T is *continuous at zero* if for every net (f_i) in F, $\|f_i\| \to 0$ implies $\|Tf_i\| \to 0$; *continuous for order* if $f_i \downarrow f$ implies that $Tf_i \downarrow Tf$; *continuous for order at 0* if $f_i \downarrow 0$ implies that $Tf_i \downarrow 0$.

Lemma 1.1. *Let E be a Banach lattice with order continuous norm, F a Banach sublattice of E, and T a positively homogeneous, increasing map from F^+ to E. Then* (i) *T is continuous at zero and continuous at zero for order.* (ii) *If in addition*

T is subadditive (hence sublinear), then T is continuous for order.

Proof: (i) Let I be a directed set, and let $(f_i, i \in I)$ be a net of elements of F^+ such that $\lim \|f_i\| = 0$ and $\sup \|Tf_i\| > \epsilon > 0$. Choose a sequence (i_n) of indices such that

$$\sum_{n=1}^{\infty} 2^n \|f_{i_n}\| < \infty \text{ and } \inf \|Tf_{i_n}\| \geq \epsilon.$$

Set

$$g_n = \sum_{k=1}^{n} 2^k f_{i_k}.$$

Since F is closed, $g_n \uparrow g \in F^+$, and for every n,

$$Tg \geq Tg_n \geq T(2^n f_{i_n}) = 2^n T(f_{i_n}),$$

hence

$$\|Tg\| \geq 2^n \epsilon.$$

This is a contradiction, therefore T must be continuous at zero. If $f_i \downarrow 0$ then $\|f_i\| \downarrow 0$, because E is order continuous. Therefore $\|Tf_i\| \downarrow 0$. Also $Tf_i \downarrow g$ for some $g \in E^+$, and necessarily $\|g\| \leq \lim \|Tf_i\| = 0$. This proves (i).

(ii) Let (f_i) be a net in F^+ such that $f_i \uparrow f$ or $f_i \downarrow f$. For every index i,

$$Tf_i \leq Tf \leq Tf_i + T(f - f_i) \text{ or } Tf \leq Tf_i \leq Tf + T(f_i - f).$$

In either case, $|Tf - Tf_i| \leq T|f - f_i|$. Now, by the order-continuity of the norm, $\lim \|f_i - f\| = 0$, hence, by part (i), $\lim \|T|f_i - f|\| = 0$ and $\lim \||Tf - Tf_i|\| = 0$. Therefore $|Tf - Tf_i| \downarrow 0$. ∎

The following propositions will be useful in deducing multi-parameter theorems from their one-parameter counterparts.

Proposition 1.2. *Let E be a Banach lattice with an order-continuous norm, F a Banach sublattice of E. Let I, J be directed sets with countable cofinal subsets. Let $(V_i, i \in I)$ be a net of increasing, sublinear maps, $V_i : F^+ \to E$.*

(i) *Assume that there is an increasing, sublinear map* $V_\infty : F^+ \to E$ *such that for each* $f \in F^+$

$$\liminf_i V_i f \geq V_\infty f.$$

Let $(f_j : j \in J)$ *be a net of elements of* F^+ *such that*

$$\liminf_j f_j = f_\infty \in F^+.$$

Then

$$\liminf_{i,j} V_i f_j \geq V_\infty f_\infty.$$

(ii) *Assume that for each* $f \in F^+$, $V_\infty f = \limsup_i V_i f \in E$. *Let* $(f_j, j \in J)$ *be a net of elements of* F^+ *such that* $\sup f_j \in F^+$. *Let* $f_\infty = \limsup f_j$. *Then*

$$\limsup_{i,j} V_i f_j \leq V_\infty f_\infty.$$

(iii) *Assume that* $\lim V_i f = V_\infty f$ *exists in* E *for each* $f \in F^+$. *Let* $(f_j, j \in J)$ *be a net of elements of* F *such that* $\sup |f_j| \in F^+$ *and* $\lim f_j = f_\infty \in F$. *Then for each* $f \in F^+$

$$\lim_{i,j} V_i f_j = V_\infty f_\infty.$$

Proof of (i): For each $j \in J$, set $m_j = \inf_{k \geq j} f_k$. By assumption, the net $m_j, j \in J$ increases to f_∞. Now in an order continuous lattice one has for a net of positive elements f_k that $\inf_k T f_k \geq T \inf_k f_k$ (Fatou's Lemma for T). Hence, for each fixed j and each u,

$$\inf_{k \geq j} V_u f_k \geq V_u m_j.$$

Therefore, for each i and j,

$$\inf_{u \geq i} \inf_{k \geq j} V_u f_k \geq \inf_{u \geq i} V_u m_j.$$

Letting $i \to \infty$ yields

$$\liminf_i \inf_{k \geq j} V_i f_k \geq \liminf_i V_i m_j \geq V_\infty m_j.$$

For each j,

$$\liminf_i \inf_{k \geq j} V_i f_k \leq \liminf_{i,j} V_i f_j.$$

Therefore

$$\liminf_{i,j} V_i f_j \geq V_\infty m_j.$$

Now the net m_j increases to $f_\infty \in F^+$. The operator V_∞ is monotonely continuous by Lemma 1.1. Hence $V_\infty m_j \uparrow V_\infty f_\infty$. It follows that

$$\liminf_{i,j} V_i f_j \geq V_\infty f_\infty.$$

Proof of (ii): Necessarily $f_\infty \in F^+$. For each $j \in J$, let $M_j = \sup_{k \geq j} f_k$. Applying Fatou's Lemma as before, we have for each u, j,

$$\sup_{k \geq j} V_u f_k \leq V_u M_j.$$

Therefore, for each i and j,

$$\sup_{u \geq i} \sup_{k \geq j} V_u f_k \leq \sup_{u \geq i} V_u M_j.$$

Letting $i \to \infty$ yields

$$\limsup_i \sup_{k \geq j} V_i f_k \leq \limsup_i V_i M_j = V_\infty M_j.$$

For each j,

$$\limsup_i \sup_{k \geq j} V_i f_k \geq \limsup_{i,j} V_i f_j.$$

Therefore

$$\limsup_{i,j} V_i f_j \leq V_\infty M_j.$$

Now the net M_j decreases to $f_\infty \in F^+$. The operator V_∞ is increasing and sublinear being the lim sup of such operators. Therefore, by Lemma 1.1, V_∞ is continuous for order. Hence $V_\infty M_j \downarrow V_\infty M_\infty$. It follows that

$$\limsup_{i,j} V_i f_j \leq V_\infty f_\infty.$$

Proof of (iii): The operator V_∞ is positive and linear being the limit of such operators. We can consider separately the action of V_i and V_∞ on the positive and negative part of functions, since $\lim f_j^+ = f_\infty^+$, $\lim f_j^- = f_\infty^-$. Now, by parts (i) and (ii),

$$\limsup_{i,j} V_i f_j^+ \leq V_\infty f_\infty^+,$$

hence $\lim_{i,j} V_i f_j^+ = V_\infty f_\infty^+$, and similarly $\lim_{i,j} V_i f_j^- = V_\infty f_\infty^-$. Hence $\lim_{i,j} V_i f_j = V_\infty f_\infty$. ∎

We now consider more than two parameters. For each $d \in N$, let $I^d = I_1 \times I_2 \times \cdots \times I_d$, where $I_k = I$ is a directed set with a countable cofinal subset. The partial order on I^d is defined by $s = (s_1, \ldots, s_d) \leq t = (t_1, \ldots, t_d)$ iff $s_k \leq t_k$ for $k = 1, \ldots, d$. The notation $t \to \infty$ then means that all the indices t_i converge to infinity independently. $L(0)$ has been introduced below to allow for a compact description of the action of the operators $T(i,j)$.

Theorem 1.3. *Let $L(0) = L(1) \supset L(2) \supset \ldots \supset L(d)$ be Banach lattices with order continuous norms and let I_i, $1 \leq i \leq d$ be directed sets with countable cofinal subsets. For each $i = 1, 2, \ldots, d$, let $(T(i,j), j \in I_i)$ be a net of increasing, sublinear maps $T(i,j) : L(i)^+ \to L(i-1)^+$.*

(i) *Assume that for each $i = 1, \ldots, d$, there exists an increasing, sublinear operator $T(i,\infty) : L(i)^+ \to L(i-1)^+$ such that for every $f \in L(i)^+$*

$$T(i,\infty) f \leq \liminf_j T(i,j) f \in L(i)^+.$$

Then for every $f \in L(d)^+$ *we have*

$$\liminf_t T(1,t_1)T(2,t_2)\cdots T(d,t_d)f \geq T(1,\infty)\cdots T(d,\infty)f.$$

(ii) *Assume that*

$$\text{(a)} \qquad \limsup_j T(i,j)f = T(i,\infty)f \in L(1)$$

for $1 \leq i \leq d$ *and each* $f \in L(1)^+$, *and*

$$\text{(b)} \qquad \sup_j T(i,j)f \in L(i-1)$$

for each $f \in L(i)^+$, $2 \leq i \leq d$. *Then for each* $f \in L(d)^+$

$$\limsup_t T(1,t_1)T(2,t_2)\cdots T(d,t_d)f \leq T(1,\infty)\cdots T(d,\infty)f.$$

(iii) *Assume that*

$$\text{(a)} \qquad \lim_j T(i,j)f = T(i,\infty)f$$

exists and is in $L(1)$ *for each* $f \in L(i)^+$ *and*

$$\text{(b)} \qquad \sup_j T(i,j)f \in L(i-1)$$

for each $f \in L(i)^+$, $2 \leq i \leq d$. *Then for each* $f \in L(d)^+$

$$\lim_t T(1,t_1)T(2,t_2)\cdots T(d,t_d)f = T(1,\infty)\cdots T(d,\infty)f.$$

Proof of (i): By induction on d. For $d = 2$ choose $f \in L(2)^+$ and apply Proposition 1.2 with $F = L(2)$, $E = L(1)$, $J = I_2$, $f_j = T(2,j)f$, $f_\infty = \liminf_j f_j$, $I = I_1$, $V_i = T(1,i)$, $V_\infty = T(1,\infty)$. Then $\liminf_i V_i f \geq V_\infty f$ by assumption. It follows that

$$\liminf_{i,j} T(1,i)T(2,j)f = \liminf_{i,j} V_i f_j \geq V_\infty f_\infty =$$

$$= T(1,\infty)\liminf_j T(2,j)f \geq T(1,\infty)T(2,\infty)f.$$

Now suppose that the inequality holds for any product of d operators. Let $F = L(d+1)$, $E = L(d)$, $f \in L(d+1)^+$, $J = I_{d+1}$, $f_j = T(d+1,j)f$, $f_\infty = \liminf_j f_j$, $I = I_1 \times \ldots \times I_d$. For $i = (i_1,\ldots,i_d) \in I$, set $V_i = T(1,i_1)\cdots T(1,i_d)$, $V_\infty = T(1,\infty)\cdots T(d,\infty)$. Since each map $T(i,\infty)$ is increasing and sublinear, the map V_∞ has the same properties. Applying Proposition 1.2 and the induction hypothesis we get

$$\liminf_i T(1,i_1)\cdots T(d,i_d)T(d+1,i_{d+1})f = \liminf_{i,j} V_i T(d+1,j)f \geq$$

$$\geq V_\infty f_\infty \geq V_\infty T(d+1,\infty)f = T(1,\infty)\cdots T(d,\infty)T(d+1,\infty)f.$$

Proof of (ii): By induction on d. For $d = 2$, choose $f \in L(2)^+$ and apply Proposition 1.2 with $F = L(2)$, $E = L(1)$, $J = I_2$, $f_j = T(2,j)f$, $f_\infty = \limsup_j f_j$, $I = I_1$, $V_i = T(1,i)$, $V_\infty = \limsup_i T(1,i)$. Then $\liminf_i V_i f \geq V_\infty f$ by assumption. It follows that

$$\limsup_{i,j} T(1,i)T(2,j)f = \limsup_{i,j} V_i f_j \geq V_\infty f_\infty =$$

$$T(1,\infty)\limsup_j T(2,j)f = T(1,\infty)T(2,\infty)f.$$

Now suppose that the inequality holds for any product of d operators. Let $F = L(d+1)$, $E = L(d)$, $f \in L(d+1)^+$, $J = I_{d+1}$, $f_j = T(d+1,j)f$, $f_\infty = \limsup_j f_j$, $I = I_1 \times \ldots \times I_d$. For $i = (i_1,\ldots,i_d) \in I$, set $V_i = T(1,i_1)\cdots T(1,i_d)$, $V_\infty = T(1,\infty)\cdots T(d,\infty)$. Since each map $T(i,\infty)$ is increasing and sublinear, so is the map V_∞. By assumption $\sup_j T(d+1,j)f \in L(d)$. Applying Proposition 1.2 and the induction hypothesis

$$\limsup_i T(1,i_1)\cdots T(d,i_d)T(d+1,i_{d+1})f = \limsup_{i,j} V_i T(d+1,j)f \leq$$

$$\leq \limsup_i V_i f_\infty = V_\infty f_\infty = T(1,\infty)\cdots T(d,\infty)T(d+1,\infty)f.$$

Proof of (iii): By parts (i) and (ii),

$$\limsup_i T(1,i_1)\cdots T(d,i_d)f \le T(1,\infty)\cdots T(d,\infty)f \le$$

$$\le \liminf_i T(1,i_1)\cdots T(d,i_d)f.$$

Hence the theorem follows. ∎

2. Convergence of multiparameter martingales

We first review the needed results about maximal inequalities and function spaces. Recall that an Orlicz function is an increasing convex function $\Phi : [0,\infty) \to [0,\infty)$, satisfying $\Phi(0) = 0$ and $\Phi(u) > 0$ for some u. Such a Φ is differentiable a.e.; the derivative φ, defined a.e. by

$$\Phi(u) = \int_0^u \varphi(x)dx$$

will be assumed left-continuous. Note that φ is increasing because Φ is convex. It follows that for every $x, y \ge 0$,

$$\Phi(x+y) \ge \Phi(x) + \Phi(y). \tag{2a}$$

We assume in addition that $\Phi(u)/u \to \infty$, which happens if and only if φ is unbounded. Then there is a "generalized left-continuous inverse" ψ of φ defined by

$$\psi(y) = \inf\{x \in (0,\infty) : \varphi(x) \ge y\}.$$

The function ψ is the derivative of an Orlicz function Ψ, called the *conjugate* of Φ. We are also interested in the function ξ defined by

$$\xi(u) = u\varphi(u) - \Phi(u) = \Psi(\varphi(u)),$$

which we call the *associate* of Φ. This associate ξ is left-continuous; it may or may not be an Orlicz function.

Let $(\Omega, \mathcal{F}, \mu)$ be a probability space. The *Orlicz modular* for Φ is the function

$$M_\Phi(f) = \int \Phi(|f|)d\mu.$$

The *Luxemburg norm* of a measurable function f is

$$\|f\|_\Phi = \inf\{a > 0 : M_\Phi(f/a) < 1\}.$$

If Φ is an Orlicz function then the space

$$L_\Phi = \{f : \|f\|_\Phi < \infty\} = \{f : M_\Phi(f/a) < \infty \text{ for some } a > 0\}$$

is a Banach lattice, called the *Orlicz space* for Φ.

The important particular cases are:

1. The L_p spaces, $1 < p < \infty$. Then $\Phi(u) = u^p/p$ and $L_\Phi = L_p$. In this case $\xi(u) = (p-1)\Phi(u)$ and L_ξ is again L_p. These spaces have an order-continuous norm.

2. The $L\log^k L$ spaces. Then $\Phi_k(u) = 0$ if $0 \leq u \leq 1$ and $\Phi_k(u) = u(\log^+ u)^k$ if $u > 1$. Φ_k is an Orlicz function if $k \geq 1$. Here $\xi(u) = k\Phi_{k-1}(u)$, hence $L_\xi = L\log^{k-1} L$. Since the measure space is finite, the $L\log^k L$ spaces have an order continuous norm (see e.g. [**3**]).

In applications of Theorem 1.3 below, the space $L(0) = L(1)$ will be L_1, the spaces $L(k)$ will be $L\log^k L$. The operator $T(i,j)$ will be E_j^i. It will be necessary to show that if a function f is in $L\log^k L$, then the appropriate supremum, called g, is in $L\log^{k-1} L$. This is done by showing that a "weak maximal inequality" implies a "strong maximal inequality". The strong inequality does not appear explicitly in the following lemma (see e.g. [**29**]), which, however, is sufficient for our purposes.

Lemma 2.1. *Let f,g be positive functions such that for some $c > 0$ and every $\lambda > 0$,*

$$\lambda\mu(\{g > c\lambda\}) \leq \int_{\{f > t\}} f \, d\mu. \tag{2.1a}$$

Then $f \in L\log^k L$ *implies* $g \in L\log^{k-1} L$.

Note that a more general conclusion about f and g belonging to associated pairs of function spaces is also true: the inequality (2.1a) implies that for any Orlicz function Φ, if $f \in L_\Phi$ then $g \in L_\xi$;cf. [**13**], Corollary 1, p. 119.

For conditional expectations, the inequality (2.1a) is classic. We state it as

Lemma 2.2. *Let* $f_n = E[f|\mathcal{F}_n]$, $f \in L_1^+$, $g = \sup f_n$. *Then*

$$\lambda\mu(\{g > \lambda\}) \leq \int_{\{f>t\}} f \, d\mu. \tag{2.2a}$$

Therefore $f \in L\log^k L$ *implies that* $g \in L\log^{k-1} L$.

We now apply these results to martingales in several parameters. Let $(\Omega, \mathcal{F}, \mu)$ be a probability space. Fix a positive integer d. Given d copies $N_1, \ldots, N_d$ of the set of positive integers N, define the directed set $I = N_1 \times \cdots \times N_d$, where the ordering on I is given by $s = (s_1, ..., s_d) \leq t = (t_1, \ldots, t_d)$ if and only if $s_i \leq t_i$ for all i, $1 \leq i \leq d$. Let $(\mathcal{F}_t, t \in I)$ be a net of sub-σ-algebras of $\mathcal{F}$ such that if $s \leq t$ then $\mathcal{F}_s \subseteq \mathcal{F}_t$. For integers i, j with $1 \leq i \leq j \leq d$, denote by $\mathcal{F}_t^{i--j}$ the σ-algebra obtained by lumping together the σ-algebras on all the axes except for the i-th, $i+1$st, ..., jth. More precisely,

$$\mathcal{F}_s^{i--j} = \bigvee_{s_k, k=1,\ldots,i-1,j+1,\ldots,d} \mathcal{F}_s.$$

If $i = j$, then we may write $\mathcal{F}_s^i$ or $\mathcal{F}_{s_i}^i$ for $\mathcal{F}_s^{i--j}$. Denote the conditional expectations $E[\,\cdot\,|\mathcal{F}_s^{i--j}]$ by E_s^{i--j} and $E[\,\cdot\,|\mathcal{F}_s]$ by E_s.

An integrable process (X_t) is called a *martingale (submartingale)* if $X_s = (\leq)$ $E_s(X_t)$ for $s \leq t$; it is a *block k-martingale (block k-submartingale)* for a fixed k if

$$X_{(s_1,\ldots,s_k,t_{k+1},\ldots,t_d)} = (\leq) E_s^{1--k}(X_t)$$

for $s \leq t$. An integrable process is called a *block martingale (block submartingale)* if it is a block k-martingale (block k-submartingale) for all k, $1 \leq k \leq d$.

It is easy to see that under the independence assumption of Cairoli [**4**], every martingale is a block martingale. It will be shown that block martingales can be factored, so convergence theorems for them reduce to those of Section 1.

We use below that on directed sets stochastic limits of L_1-bounded martingales and submartingales exist, and L_1 limits of martingales given by conditional expectations (uniformly integrable martingales) exist. Indeed, stochastic convergence and convergence in L_1 are defined by a metric, so behavior of sequences is determining (see e.g. [**26**],p. 96). Stochastic limit is denoted by $s\lim$.

Proposition 2.3. *Let J be a directed set, $(\mathcal{F}_t : t \in J)$ a filtration, and Φ an Orlicz function with $\Phi(u)/u \to \infty$. Let $(X_t : t \in J)$ be a martingale or a positive submartingale such that $\Phi(|X_t|)$ is bounded in L_1. Then there is a random variable X such that $\Phi(|X_t - X|)$ converges to 0 in L_1.*

Proof: Let $s\lim X_t = X$. Since X_t is bounded in L_Φ, it is uniformly integrable and converges to X in L_1. Assume X_t is a positive submartingale. Then $\Phi(X_t)$ converges stochastically to $\Phi(X)$ and by Fatou's lemma, $\Phi(X) \in L_1$. If $t \leq u$ and $A \in \mathcal{F}_t$ then $\int_A X_t \, d\mu \leq \int_A X_u \, d\mu$ by the submartingale property. Taking the limit as $u \to \infty$ we get $\int_A X_t \, d\mu \leq \int_A X \, d\mu$ for all $A \in \mathcal{F}_t$. Hence $X_t \leq E_t X$. Now, by Jensen's inequality, we have $\Phi(X_t) \leq E_t \Phi(X)$ which is uniformly integrable. It follows that $\Phi(X_t)$ is uniformly integrable. Hence $\Phi(X_t)$ converges to $\Phi(X)$ in L_1. Note that by (2a) we have

$$\Phi(|X_t - X|) \leq |\Phi(X_t) - \Phi(X)|,$$

hence $\Phi(|X_t - X|)$ converges to zero in L_1.

The convergence of a martingale X_t follows because $X_t = X_t^+ - X_t^-$ and both X_t^+ and X_t^- are submartingales. ∎

The following is a characterization of block martingales.

Proposition 2.4. *For a uniformly integrable martingale (X_t) with $X = s\lim X_t$, the following are equivalent: (a) $X_t = E_t X$ is a block martingale; (b) for each $k \leq d$ and each t*

$$E_t X = E_t^{1--k} E_t^{(k+1)--d} X. \tag{2.4a}$$

If (X_t) is a uniformly integrable block submartingale with $X = s\lim X_t$ then (2.4a) holds with the inequality $\leq$ replacing the equality.

Proof: (a) implies (b). Let $s = (s_1, \ldots, s_d) \leq t = (t_1, \ldots, t_d)$, $t_{k+1} = s_{k+1}, \ldots, t_d = s_d$. X_t is a block-k-martingale, therefore we have

$$X_s = X_{(s_1,\ldots,s_k,t_{k+1},\ldots,t_d)} = E_s^{1--k} X_t.$$

Taking the limit in L_1 when $t_1 \to \infty, \ldots, t_k \to \infty$, we have that $\lim X_t = E_s^{(k+1)--d} X$, hence

$$X_s = E_s^{1--k} E_s^{(k+1)--d} X.$$

The argument is the same for block submartingales, with the inequality $\leq$ replacing the equality.

(b) implies (a). Assume (2.4a) holds for each k and let $s \leq t$. Then

$$E_s^{1--k} X_t = E_s^{1--k} E_t^{1--k} E_t^{(k+1)--d} X =$$

$$= E_s^{1--k} E_t^{(k+1)--d} X = E_s^{1--k} E_s^{(k+1)--d} X = X_s.$$

∎

Theorem 2.5. *Let X_t be a uniformly integrable block martingale, $X = s\lim X_t$. Then for all s,*

$$X_s = E_s^1 E_s^2 \cdots E_s^d X. \tag{2.5a}$$

(ii) If X_t is a uniformly integrable block submartingale, then (2.5a) holds with the inequality $\leq$ replacing the equality.

Proof: (i) We are going to show by induction on k that for all k

$$X_s = E_s^1 \cdots E_s^k E_s^{(k+1)--d} X. \tag{2.5b}$$

Case $k = 1$. The equality

$$X_s = E_s^1 E_s^{2--d} X$$

follows from Proposition 2.4.

Now let us assume that (2.5b) holds for k. We have, by Proposition 2.4,

$$X_s = E_s^{1--(k+1)} E_s^{(k+2)--d} X.$$

Let $s_1 \to \infty, \ldots, s_k \to \infty$. It follows that

$$E_s^{(k+1)--d} X = E_s^{(k+1)} E_s^{(k+2)--d} X.$$

We substitute this into (2.5b) to complete the induction step.

(ii) The proof is similar to that of (i). ∎

Theorem 2.6. *(i) Let (X_t) be a block submartingale bounded in $L \log^{d-1} L$. Denote by X the stochastic limit of X_t. Then* $\limsup X_t = X$ *(upper demiconvergence).*
(ii) Let (X_t) be a block martingale bounded in $L \log^{d-1} L$. Then X_t converges almost everywhere and in $L \log^{d-1} L$ to a random variable X and $X_t = E_t X$.

Proof: (i) Assume first that (X_t) is a positive block submartingale. Apply Theorem 1.3 (ii) and Lemma 2.2 with $T(i,t) = E_t^i$, $L(i) = L \log^{i-1} L$, $1 \le i \le d$ and $g = \sup_t E_t^i X$. Note that $E_t^1 X$ converges a.e. as $t_1 \to \infty$ to $E[X | \vee_{t_1} \mathcal{F}_{t_1}^1] = X$, hence $T(1,\infty)X = X$ a.e. Similarly $T(i,\infty)X = X$ a.e. for all i. Hence by Theorem 2.5 and Theorem 1.3 (ii),

$$\limsup X_t \le \limsup E_t^1 E_t^2 \cdots E_t^d X \le T(1,\infty) \cdots T(d,\infty) X = X.$$

On the other hand we always have $X = s\lim X_t \le \limsup X_t$. Hence the statement follows for positive block submartingales, hence for block submartingales bounded from below by a constant. Now let X_t be arbitrary. For a fixed real number a, the process $(X_t \vee a)$ is also a block submartingale, hence

$$\limsup(X_t \vee a) = X \vee a$$

by the first part. Note that $\limsup X_t > -\infty$ by Fatou's lemma. Since a was arbitrary, it follows that $\limsup X_t = X$.

(ii) If X_t is a block martingale then X_t and $-X_t$ are block submartingales. Thus $\limsup X_t = X$ and $-\liminf X_t = \limsup(-X_t) = -X$. ∎

Block martingales and submartingales, and therefore the results of this section are applicable to averages of exchangeable random variables in several parameters. Thus in [**12**] it is shown that if $\mu(\Omega) = 1$, $Y_{ij} \in L_p^+$ and the random variables Y_{ij} are exchangeable (actually only "T-exchangeability" is needed) then the random variables

$$X_t = X_{-m,-n} = \frac{1}{mn}(\sum_{i=1}^{m}\sum_{j=1}^{n} Y_{ij})^p$$

form a reversed 1-submartingale (supermartingale) if $p \leq 1$ ($p \geq 1$). The theory of reversed martingales, submartingales, etc., is similar to the theory of direct processes, but rather easier: representations of one-parameter martingales as conditional expectations is automatic. One parameter theorems hold, and the principles of Section 1 apply without change. Almost everywhere convergence follows if $p \leq 1$. For $p < 1$, the one-parameter case is due to Marcinkiewicz.

3. A multiparameter ratio ergodic theorem

Let T be a positive contraction on L_1. Then the space Ω decomposes into the conservative part C and the dissipative part D. For each $f \in L_1^+$, $\sum_{i=0}^{\infty} T^i f = 0$ or ∞ on C, $\sum_{i=0}^{\infty} T^i f < \infty$ on D (Hopf's decomposition, see e.g. [**21**]). T is called Markovian if it preserves the integral: $\int Tf = \int f$ for each $f \in L_1$.

A sequence (s_n) of functions in L_1^+ is called a *superadditive*, if

$$s_{k+n} \geq s_k + T^k s_n \tag{3a}$$

for every $k, n \geq 0$, and

$$\gamma = \sup_n \frac{1}{n} \int s_n d\mu < \infty.$$

The constant γ is called the *time constant* of the process. The sequence (s_n) is *subadditive* if $(-s_n)$ is superadditive; (s_n) is *additive* if it is both superadditive and subadditive. If only assumption (3a) is made, the sequence (s_n) is called *extended superadditive.*

Note that by (3a), we have for all n,

$$s_n \geq \sum_0^{n-1} T^k s_1. \tag{3b}$$

There is a superadditive ratio theorem that generalizes both Kingman's theorem [**19**] and (for positive operators) Chacon's ratio theorem [**6**]. We state it here as

Theorem 3.1. [2] *Let T be a positive contraction on L_1. Suppose (r_n) is a positive superadditive sequence and (s_n) is a positive extended superadditive sequence, $E = \{\sup_n s_n > 0\}$. Then the ratio r_n/s_n converges a.e. to a finite limit on the set $C \cap E$. If either T is Markovian or (r_n) is additive on D, then* $\lim(r_n/s_n) = (\lim \uparrow r_n)/(\lim \uparrow s_n) < \infty$ *exists a.e. on $D \cap E$.*

We are going to use Theorem 1.3 to prove a multiparameter variant of this result, in which the sequence in the numerator is additive. The case where both the numerator and the denominator are additive was proved in [**16**].

Assume $s_1 > 0$ and let $f \in L_1^+$. Define

$$h = \sup_n \frac{\sum_0^n T^i f}{s_{n+1}},$$

and let

$$g = \sup_n \frac{\sum_0^n T^i f}{\sum_0^n T^i s_1}.$$

Lemma 3.2. *For every $\lambda > 0$,*

$$\lambda \int_{\{g>\lambda\}} s_1 d\mu \leq \int_{\{g>\lambda\}} f d\mu. \tag{3.2a}$$

Proof: We have

$$\{g > \lambda\} = \{\sup_n \sum_0^n T^i(f - \lambda s_1) > 0\}.$$

Hence, by Hopf's Maximal Ergodic Theorem, (cf. [**21**], p. 8)

$$\int_{\{g>\lambda\}} (f - \lambda s_1) d\mu \geq 0,$$

which is equivalent to (3.2a). ∎

Let ν be the measure $s_1 \cdot \mu$. Then ν is a finite measure, equivalent to μ. By Lemma 3.2,

$$\begin{aligned} 2\lambda\nu(g > 2\lambda) &\leq \int_{\{g>2\lambda\}} f\, d\mu = \int_{\{g>2\lambda\}} f/s_1 d\nu \leq \\ &\leq \int_{\{f/s_1>\lambda\}} f/s_1 d\nu + \int_{\{f/s_1<\lambda, g>2\lambda\}} f/s_1 d\nu \leq \\ &\leq \int_{\{f/s_1>\lambda\}} f/s_1 d\nu + \lambda\nu(g > 2\lambda), \end{aligned}$$

hence

$$\nu(g > 2\lambda) \leq 1/\lambda \int_{\{f/s_1>\lambda\}} f/s_1 d\nu. \tag{3.2b}$$

Lemma 3.3. *If $f/s_1 \in L \log^k L(\nu)$ then $h \in L \log^{k-1} L(\nu)$.*

Proof: The implication with g instead of h follows from Lemma 2.2. Since $h \leq g$, the lemma is proved. ∎

We now state the ratio theorem.

Theorem 3.4. *Let $(X, \mathcal{F}, \mu)$ be a finite measure space. Assume that T_i are positive contractions on $L_1(\mu)$, $s_n^{(i)}$ are positive extended superadditive sequences with* *respect to* T_i, $i = 1, \ldots, d$. *Let* ν_i *be the measure* $s_1^{(i)} \cdot \mu$. *Suppose the functions* $s_1^{(1)}, s_1^{(2)}/s_1^{(1)}, \ldots, s_1^{(d)}/s_1^{(d-1)}$ *are bounded away from zero. Then for each* f *such that* $f/s_1^{(d)} \in L \log^{d-1} L(\nu_d)$,

$$\frac{\sum_{k_1=0}^{n_1-1} T_1^{k_1}}{s_{n_1}^{(1)}} \cdots \frac{\sum_{k_d=0}^{n_d-1} T_d^{k_d} f}{s_{n_d}^{(d)}}$$

converges a.e. as $n_1 \to \infty, \ldots, n_d \to \infty$ *independently.*

Proof: Define the operators

$$T(k,n) = \frac{\sum_{i=0}^{n} T_k^i}{s_{n+1}^{(k)}}.$$

Let, for $k = 1, \ldots, d$,

$$L(k) = \{f : f/s_1^{(k)} \in L \log^{k-1} L(\nu_k)\}.$$

Define $L(0)$ as $L(1) = L_1(\nu_1)$. The norm of f in $L(k)$ is defined as the norm of $f/s_1^{(k)}$ in $L \log^{k-1} L(\nu_k)$. Since $L \log^{k-1} L(\nu_k)$ is a Banach lattice with order-continuous norm, so is $L(k)$. Also, if $f \in L_1(\mu)$, then $T(k,n)f \in L_1(\nu_k) \subset L_1(\nu_{k-1}) \subset \ldots \subset L_1(\nu_1) \subset L_1(\mu)$. Suppose $f \in L(k)$. Then, by Lemma 3.3,

$$h_k = \sup_n T(k,n) f \in L \log^{k-2} L(\nu_k).$$

Since $s_1^{(k)}/s_1^{(k-1)}$ is bounded away from zero, we have

$$L \log^{k-2} L(\nu_k) \subset L \log^{k-1} L(\nu_{k-2}).$$

Since the assumptions imply that $s_1^{(k-1)}$ is also bounded away from zero, $h_k \in L(k-1)$. The existence of the limits $\lim_{n\to\infty} T(k,n)f$ follows from Theorem 3.1; since (r_n) is additive, it is not necessary to assume that T is Markovian. The proof is completed by Theorem 1.3. ∎

References

1. M. A. Akcoglu, *A pointwise ergodic theorem for L_p spaces*, Canad. J. Math. **27** (1975), 1075-1082.

2. M. A. Akcoglu and L. Sucheston, *A ratio ergodic theorem for superadditive processes*, Z. Wahrscheinlichtkeitstheorie verw. Geb. **44** (1978), 269-278.

3. M. A. Akcoglu and L. Sucheston, *On uniform monotonicity of norms and ergodic theorems in function spaces*, Supplemento ai Rendiconti del Circolo Matematico di Palermo **8** (1985), 325-335.

4. R. Cairoli, *Une inégalité pour martingales à indices multiples et les applications*, Lecture Notes Math. **124**, 1-28. Berlin-Heidelberg-New York: Springer 1970.

5. R. Cairoli and J. B. Walsh, *Stochastic integrals in the plane*, Acta M. **134** (1975), 111-183.

6. R. V. Chacon, *Convergence of Operator averages*, In: Ergodic Theory, 89-120, New York: Academic Press 1963.

7. R. V. Chacon and D. S. Ornstein, *A general ergodic theorem*, Illinois J. Math. **4** (1960), 153-160.

8. C. Dellacherie and P. A. Meyer, *Probabilities and Potential*, Amsterdam: North Holland, 1982.

9. J. L. Doob, *Stochastic processes*, Wiley, New York, 1953.

10. N. Dunford, *An individual ergodic theorem for noncommutative transformations*, Acta Sci. Math. (Szeged) **14** (1951), 1-4.

11. N. Dunford and J. Schwartz, *Linear operators*, Interscience Publ., 1958.

12. G. A. Edgar and L. Sucheston, *Démonstrations de lois des grands nombres par les sous-martingales descendantes*, C.A. Acad. Sc. Paris **292** (1981), 967-969.

13. G. A. Edgar and L. Sucheston, *On Maximal Inequalities in Orlicz Spaces*, Contemporary Mathematics **94** (1989), 113-129.

14. N. A. Fava, *Weak inequalities for product operators*, Studia Math. **42** (1972), 271-288.

15. N. Frangos and L. Sucheston, *On convergence and demiconvergence of block martingales and submartingales*, Proceedings of the Fifth International Conference on Probability in Banach Spaces, Tufts University, 1984. Lecture Notes in Math. **1153** (1985), 198-225.

16. N. Frangos and L. Sucheston, *On multiparameter ergodic and martingale theorems in infinite measure spaces*, Probab. Th. Rel. Fields **71** (1986), 477-490.

17. A. and C. Ionescu Tulcea, *Abstract ergodic theorems*, Trans. Amer. Math. Soc. **187** (1963), 107-124.

18. S. Kakutani, *Ergodic theory*, Proc. International Congress of Mathematicians. **2** (1950), 129-142.

19. J. F. C. Kingman, *The ergodic theory of subadditive stochastic processes*, J. Roy. Statist. Soc. Ser. B, **30** (1968), 499-510.

20. M. A. Krasnosel'skii and Ya. B. Rutickii, *Convex Functions and Orlicz Spaces*, New York: Gordon and Breach Science Publishers 1961.

21. U. Krengel, *Ergodic theorems*, De Gruyter Studies in Mathematics **6** (1985).

22. J. Lindenstrauss and L. Tzafriri, *Classical Banach spaces II, function spaces*, Springer, 1979.

23. A. Millet and L. Sucheston, *On regularity of multiparameter amarts and martingales*, Z. Wahrscheinlichtkeitstheorie verw. Geb. **56** (1981), 21-45.

24. A. Millet and L. Sucheston, *Demiconvergence of processes indexed by two indices*, Ann. Inst. Henri Poincaré **19** (1983), 175-187.

25. A. Millet and L. Sucheston, *On fixed points and multiparameter ergodic theorems in Banach lattices*, Can J. Math. **15** (1988), 429-458.

26. J. Neveu, *Discrete-Parameter Martingales*, North-Holland Mathematical Library **10**, North-Holland, 1975.

27. G. C. Rota, *An "alternierende Verfahren" for general positive operators*, Bull. Am. Math. Soc. **68** (1962), 15-102.

28. E. M. Stein, *On the maximal ergodic theorem*, Proc. Nat. Acad. Sci. USA **47** (1961), 1894-1897.

29. L. Sucheston, *On one-parameter proofs of almost sure convergence of multiparameter processes*, Z. Wahrscheinlichtkeitstheorie verw. Geb. **63** (1984), 43-49.

30. A. Zygmund, *An individual ergodic theorem for noncommutative transformations*, Acta Sci. Math. (Szeged) **14** (1951), 103-110.